Collective Intelligence

Other Books by Pierre Lévy

Becoming Virtual: Reality in the Digital Age

Collective Intelligence
Mankind's Emerging World in Cyberspace

PIERRE LÉVY

Translated from the French by
Robert Bononno

§

HELIX BOOKS

PERSEUS BOOKS

Cambridge, Massachusetts

Library of Congress Catalog Card Number: 99-066855

ISBN 0-7382-0261-4

Perseus Books is a member of the Perseus Books Group.

1 2 3 4 5 6 7 8 9 10——03 02 01 00 99
First printing, November 1999

Find Helix Books on the World Wide Web at
http://www.perseusbooks.com

To Raphael Cohen and Bill Dunn

Foreword

Pierre Lévy's *Collective Intelligence: Mankind's Emerging World in Cyberspace* is an important contribution to the field of computers and social theory. The following translation (from its original French) by Robert Bononno brings to the attention of American and English-speaking readers one of the Continent's most imaginative thinkers about computers and their impact on society and culture.

Lévy's work is part of a long line of inquiry by French social theorists into technology and communication that extends back to Jacque Ellul's *The Technological Society*[1] and Simon Nora and Alain Minc's *The Computerization of Society*.[2] Its metaphysical and epistemological roots also recall the work of the French religious philosopher Pierre Teilhard de Chardin.[3]

Thinking about computers the way Lévy does in *Collective Intelligence* is atypical of most Americans writing about computers. His work is not only deeply cultural and philosophical, but in the end, is also deeply metaphysical and utopian. Lévy perceives the computerization of society as part of a process of evolution that is leading us "toward the creation of a new medium of communication, thought, and work for human societies." (p. 1) For Lévy, the computer, and its associated communication technologies, is creating a "nomadic" culture in which cognitive prostheses "are transforming our intellectual capabilities as clearly as the mutations of our genetic heritage." (p. 5) Something as seemingly simple as a spell-checker on a word processing system makes us "perfect spellers." Likewise a grammar checking system renders our verb subject agreements perfectly, just as a calculator makes our mathematical computations absolutely exact.

Even more important for Lévy, the computer makes possible a shared or collective intelligence. This idea is not new, having been suggested in the 1960s by the communication theorist and popular culture figure, Marshall McLuhan, who suggested that communication technologies were in the process of creating a global village, as well as by the computing pioneer, Douglas Englebart.[4] Engelbart's significance as a thinker and innovator has only begun to be widely recognized in recent years. In fact, many of the arguments that Lévy makes in *Collective Intelligence* echo ideas developed by Engelbart. In the early 1960s, for example, Engelbart argued that computers were not simply efficiency and productivity tools, but could actually be used to "augment intellect."[5]

By combining the computational and organizational capability of the computer with human intelligence, Engelbart proposed the emergence of a way of life "in an integrated domain" where "hunches, cut-and-try, intangibles, and the human 'feel for a situation'" would "coexist with powerful concepts, streamlined terminology and notation, sophisticated methods, and high-powered electronic aids."[6] Engelbart's current work on "Collaboration," "Knowledge Management," "Virtual Teaming," and "Continuous Improvement and Learning," as part of the Bootstrap Institute suggests similar lines of thought to those of Lévy and his concept of "collective intelligence."

How then, American readers might ask, is Lévy an original thinker? Part of the answer lies in the fact that Engelbart is coming from an engineering and business tradition. He is addressing technological problems largely through the creation and manipulation of hardware and software. Lévy's starting point is quite different. Although actively engaged in the development of innovative software, his work is more consciously grounded in social, economic, and political structures. For Lévy, "The prosperity of a nation, geographical region, business, or individual depends on their ability to navigate the knowledge space." (p. 10)

This is how Lévy's contribution is truly original. As he explains, the *knowledge space* that is being formed today will almost certainly "take precedence over the spaces of earth, territory and commerce that preceded it." (p. 13) In *Collective Intelligence*, Lévy is attempting to map or create "a detailed cartography of these spaces and their inter-relationship." (p. 13) He believes, like Engelbart, that we can develop sophisticated systems of networked intelligence. These systems—what Lévy refers to as *collective intelligence*—have the potential to project humanity into a new phase of its intellectual and social evolution.

Referring back to an earlier work co-authored with Michel Authier, Lévy argues that the creation of the new *knowledge space*, and with it the possibilities of collective intelligence, is a direct result of new computer technologies such as hypertext. According to him, the sum of knowledge is now organized by the cosmos and not the circle: "instead of a one-dimensional text or even a hyper-text network, we now have a dynamic and interactive multidimensional representational space." (p. 174) Lévy refers to this new knowledge space as the *cosmopedia*.

The *cosmopedia*, as the knowledge space around which the new *collective intelligence* is organized goes beyond the image and text characteristic of print-based encyclopedias. Instead, it combines "static images, video, sound, interactive simulation, interactive maps, expert systems, dynamic ideographs, virtual reality, artificial life, etc." (pp. 174–175) Taken to its most extreme form, the "*cosmopedia* contains as many semiotics as exist in the world itself." (p. 175)

For Lévy, the *cosmopedia* dematerializes the artificial boundaries between disciplines, making knowledge "a large patchwork" in which virtually any field can be folded onto another. The power of disciplinary knowledge is dissolved. A shared discourse and *collective intelligence* now becomes possible.

Lévy's book is essentially a utopian tract—arguably the first full-scale utopian text of the computer era. In reading *Collective Intelligence*, I found myself thinking about earlier utopian works such as Thomas More's *Utopia* and Samuel Butler's *Erewhon*. It is in this sense, that Lévy's work stands apart from Engelbart's. While Engelbart suggests utopian possibilities, Lévy's work outlines a full-blown utopian system based on the new models of computing that individuals such as Engelbart have been creating over the course of the last generation.

What is Lévy's message? He sees the computerization of society as having the potential to "promote the construction of intelligent communities in which our social and cognitive potential can be mutually developed and enhanced." (p. 17) It is his hope that new computer technologies such as the Internet and the World Wide Web "will serve to filter and help us navigate knowledge, and enable us to think collectively rather than simply haul masses of information around with us." (*Ibid*)

Lévy's vision is deeply human and social. He believes that the computer can, through technologies such as "knowledge trees," provide us a means by which to share knowledge with others and meet them in a largely unbiased and democratic cyberspace. Lévy's "collective intelligence" is a "universally distributed intelligence." He believes that "no one knows everything, everyone knows something, all knowledge resides in humanity." (p. 20) All of us have something to contribute to his information utopia. All have something to gain.

What Lévy proposes is a project that implies a new humanism "that incorporates and enlarges the scope of self knowledge into a form of group knowledge and collective thought." (p. 23) He argues that we are passing from a Cartesian model of thought based upon the singular idea of *cogito* (I think) to a collective or plural *cogitamus* (we think). The computer is the instrument that makes this utopian ideal possible.

I have argued in several of my own books that we are at a turning point in our history and culture: we are moving from a typographic or modern culture into a post-typographic or post-modern culture.[8] The parallels to the transition from Medieval to Renaissance culture—from a largely oral and spoken culture to a text and print-based culture seem clear. Just as Renaissance humanists, such as Thomas More, saw the possibilities for a new age and a new model of humanity emerging in the early phases of the Renaissance, I think that we can see the possibilities for the emergence of a new model of humanity and culture emerging in our post-modern and computer-based society. Pierre Lévy provides us with a map of this new intellectual and social terrain. His belief that new computer-based technologies such as the Internet, the World Wide Web, hypertext, and hypermedia can provide us with a new architecture for our thought in the form of *collective intelligence* is an important and profound insight. It will be interesting to see if its reality is as positive a vision as he hopes.

I am convinced that this book will infuriate many people—particularly those programmers and technicians who dominate the computer field and who have failed to grasp the cultural and social significance of computing. Lévy's contribution in *Collective Intelligence* is to provide an intriguing blueprint for how the computer and its associated technologies can expand our intelligence and humanity. He makes no claims to predict exactly what will happen in the near future. Instead, he argues that a redefinition of culture and knowledge "has begun and we do not yet know, within the context of its overall movement, what limits it will shift or how far it will shift them." (p. 201) For him, *collective intelligence* "is a utopia of the unstable and the multiple." (p. 202) As new technologies such as hypertext, hypermedia, the Internet, and the World Wide Web continue to evolve, we as a culture and society will see this new "utopia" emerge. I believe Lévy's *Collective Intelligence* will stand as an important guide to its outlines and terrain.

Eugene F. Provenzo, Jr.
Miami, Florida

Contents

II.   THE KNOWLEDGE SPACE

Prologue

The Nomad Planet

The launch of the information highway was met with considerable fanfare when first announced. The shake-ups accompanying the series of mergers, buyouts, and strategic alliances in the communications and information technology sector, the various announcements concerning the arrival of high-definition digital television, all served as signals that focused public attention on what is ordinarily referred to as multimedia.

The events that have generated so much interest in this field are individual manifestations of a great wave of technological development. Data, text, graphics, sound, messages of all types are being digitized and, to an increasing extent, produced directly in digital format. The tools for automatically processing such information have become increasingly commonplace throughout all sectors of human activity. The interconnection of computers and storage systems over standard telephone lines and the extension of digital transmission networks continue to expand a global cyberspace in which elements of information are in virtual contact with one another and with anyone who happens to be connected. The effect of these fundamental tendencies, which have already been at work for more than twenty-five years, will increase during the coming decades. The process of evolution now at work is converging toward the creation of a new medium of communication, thought, and work for human societies.

Beginning in the sixties, pioneers such as D. Engelbart and J.C.R. Licklider had perceived the social potential of computer-mediated communication. But it wasn't until the start of the eighties that digitized communication—or telematics—emerged as an economic and cultural phenomenon in its own right, comprising global networks of university students and researchers, corporate networks, electronic mail, locally based "virtual communities," and direct access to database information.

By the end of the eighties, personal computers were becoming increasingly powerful and easier to use; the number of applications multiplied and continued to grow on an almost daily basis. A parallel process was also taking place in which previously isolated networks were being interconnected, coupled with exponential growth in the number of users of digitized information. An inter-network based on the "anarchic" cooperation of thousands of computer facilities throughout the world, the Internet has become the symbol of that heterogeneous and cross-border medium that we refer to as cyberspace. Every month the number of people with an e-mail address increases by five percent throughout the world. In 1994 more than 20 million people, mostly young, were connected to the Internet. Forecasts predict 100 million users by the year 2000. Through the use of digital networks, individuals have been able to exchange messages, participate in online conferences on any of thousands of different topics, access public information, make use of the computing power of machines located thousands of miles away, construct virtual worlds, develop political projects, friendships, hatreds, cooperative efforts ... an immense living encyclopedia.[1]

Network culture has not yet stabilized, its technical infrastructure is still in its infancy, its growth far from complete. It is still not too late to work together and attempt to alter the course of events. There is still room within this new space for other ways of doing things. Will the information highway and multimedia end up as a kind of "TV deluxe?" Do they represent the final victory of consumer consumption and the spectacle? Will they increase the gulf between rich and poor, the marginalized and the "connected?" This is obviously a distinct possibility. If, however, we recognize the importance of the stakes involved in due time, these new means of creation and communication could also profoundly reshape the structure of the social bond in the direction of a greater sense of community and help us resolve the problems currently facing humanity.

The fusion of telecommunications, informatics, the news media, publishing, television, film, and electronic gaming within a unified multimedia industry is the aspect of the digital revolution that journalists have consistently emphasized. It is not the only aspect, however, nor perhaps the most important. Aside from the inevitable commercial consequences, there is some urgency in pointing out the risks to civilization associated with the emergence of multimedia: new patterns of communication, regulation, and cooperation, unknown forms of intellectual language and technology, modification of our relationship to time and space, etc. The shape and content of cyberspace are still partly undetermined. In practice there is no simple form of technological or economic determinism. Governments, large corporate entities, and individual citizens will soon be faced with fundamental political and cultural choices. The future will not be based solely on the impact of such developments (for example, the effect of the electronic highway on political, economic, or cultural life) but on our willingness to shape them as well (for what purpose will we develop interactive digital communication networks?). Whether we want them to or not, technical decisions, the adoption of standards, regulations, and tariff policies will help shape the social infrastructure of sensibility, intelligence, and coordination that will lay the foundation for the global civilization of tomorrow. This book is intended to help promote that development from an anthropological perspective and forge a positive vision that might help us orient our policies, decisions, and practices within the labyrinth of a cyberspace in the process of becoming.

The development of new instruments of communication has accelerated and in a sense transcended the large-scale process of mutation taking place. We have again become nomads.

By this I am not referring to pleasure cruises, exotic vacations, or tourism. Nor to the incessant come and go of businessmen and harried travelers, crisscrossing the globe, moving from airport to airport. Neither do the portable devices of mobile computing bring us any closer to an understanding of today's nomadism. Such images of movement reflect a world of motionless travel, enclosed within the same universe of signification. The final obstacle to the voyage may be the endless race within existing commodity networks. Movement no longer means traveling from point to point on the surface of the globe, but crossing universes of problems, lived worlds, landscapes of meaning. These wanderings among the textures of humanity may intersect the well-delineated paths of the circuits of communication and transport, but the

oblique and heterogeneous navigations of the new nomads will investigate a different space. We have become
immigrants of subjectivity.

The nomadism of today reflects the continuous and
rapid transformation of scientific, technical, economic,
professional, and mental landscapes. Even if we remain
rooted to one spot, the world will change around us. Yet
we move. And the chaotic mass of our responses produces a general transformation. Doesn't such movement
require of us some rational or optimal adaptation? But
how are we to know that a given response is appropriate
to a configuration that we are presented with for the first
time, one that no one has programmed? And why would
we want to adapt (adapt to what exactly?) once we realize
that this reality is not present before us, outside us, pre-
existent, but rather the transitory consequence of something we have constructed together?

Unpredictable, risky, the situation resembles a dizzying ride down uncharted rapids. The problem is not simply that we are traveling across a terrain that is outside
technology, economy, and civilization. If it were simply a
matter of crossing from one culture to another, we would
have numerous examples to follow, historical landmarks.
Rather, we are moving from one humanity to another, a
humanity that not only remains obscure and indeterminate, but that we refuse to interrogate, that we are still
unwilling to acknowledge.

The conquest of space involves the explicit attempt to
establish human colonies on other planets, that is to say, a
radical change of habitat and environment for our species. Through the advances of biology and medicine, we
have begun to reinvent our relation to our bodies, to
reproduction, sickness, and death. We are gradually moving, perhaps without realizing it and certainly without
acknowledgment, toward a form of artificial selection
governed by genetics. The development of nanotech-

nologies capable of mass producing intelligent materials, artificial microscopic symbionts of our bodies, and computers that are more powerful than those in use today by several orders of magnitude could profoundly alter our relationship to necessity and work, and much more brutally than the various stages of automation have yet been able to accomplish. The development of digitally controlled cognitive prostheses are transforming our intellectual capabilities as clearly as the mutations of our genetic heritage. The new technologies of communication through virtual worlds have altered the formulation of the problem of the social bond. In short, hominization, the process of the emergence of the human species, is not over. In fact it seems to be sharply accelerating.

In contrast to what occurred during the birth of our species or during the first great anthropological mutation (the Neolithic period, which saw the arrival of livestock breeding, agriculture, the city, state, and writing), we have the opportunity to collectively think through our future and alter its course.

Bureaucratic hierarchies (based on static forms of writing), media monarchies (surfing the television and media systems), and international economic networks (based on the telephone and real-time technologies) can only partially mobilize and coordinate the intelligence, experience, skills, wisdom, and imagination of humanity. For this reason the development of new ways of thinking and negotiating engendered by the growth of genuine forms of *collective intelligence* becomes particularly urgent. Intellectual technologies are not just another branch of contemporary anthropological change, they are a potential critical zone, its political nexus. There is no reason to belabor this point, however, for we can't reinvent the instruments of communication and collective thought without reinventing democracy, a distributed, active, molecular democracy. Faced with the choice of turning back or moving forward, in the presence of a precarious feed-back system, humanity has a chance to reclaim its future. Not by placing its destiny in the hands of some so-called intelligent mechanism, but by systematically producing the tools that will enable it to shape itself into intelligent communities, capable of negotiating the stormy seas of change.

This new nomadism will not develop within any known geographic territory, institution, or state, but within an invisible space of understanding, knowledge, and intellectual power, within which new qualities of being and new ways of fashioning a society will flourish and mutate. Not the space of organization charts and business statistics, but the qualitative, dynamic, living space of a humanity in the process of inventing itself through the creation of its world.

Where can we study the inconstant mass of this fluctuating space? *Terra incognita*. Even if we manage to achieve a condition of personal immobility, the landscape will continue to flow and tumble around us, infiltrate us, transform us from within. We are no longer in historical time, with its references to writing, the city, the past, but within a moving and paradoxical space that comes to us from the future. Only by means of some dangerous optical illusion will we be able to contemplate this time as a succession of events or question its traditions. Time now is errant, oblique, plural, indeterminate, like that which precedes all origins.

Crowds of refugees marching toward unlikely campsites. Nations without permanent residence. Epidemics of civil war. The clamorous babel of global metropolises. The movement of populations by means of survival skills across the interstices of empire. The impossibility of founding cities, the impossibility of establishing anything on the basis of a secret, a power, a soil, no matter where. Signs have become migrant. The earth continuously trem-

bles and burns. We slip and slide dizzyingly among religions and languages, *zap* between voices and chants,
when suddenly, near a detour around a subterranean
passage, bursts forth the music of the future. The earth,
like a marble beneath the giant eye of a satellite ...

The original nomads followed the herds, which were
themselves in search of food, according to the seasons
and the rains. Today, we continue to wander behind our
human future, a future that intersects us as we create it.
Humanity has become its own climate, an infinite season
from which there is no return. Humankind and animal
together, increasingly indistinguishable from our tools
and from a world tightly bound to our advance. With
each new day, we unfold a new landscape. Neanderthal
man, well suited for the hunt over the icy tundra, became
extinct when the climate suddenly grew warmer and
more humid.[2] Their customary quarry disappeared. In
spite of their intelligence, our grumbling or mute ancestors were without a voice, they lacked a language to
communicate with one another. Though sporadic solutions to new problems were found, they couldn't be generalized. Mankind scattered in the face of the transformation of the world taking place around him. He was unable
to change with it.

Today, *Homo sapiens* is faced with a rapid modification of his environment, a transformation for which he is
the involuntary collective agent. I am not implying that
our species is threatened with extinction or that the "end
of the world" is approaching. I am not preaching mille-
narianism. Rather, I would like to point out an alternative.
Either we cross a new threshold, enter a new stage of
hominization, by inventing some human attribute that is
as essential as language but operates at a much higher
level, or we continue to "communicate" through the media and think within the context of separate institutions,
which contribute to the suffocation and division of intel-

ligence. In the latter case we will no longer be confronted only by problems of power and survival. But if we are committed to the process of collective intelligence, we will gradually create the technologies, sign systems, forms of social organization and regulation that enable us to think as a group, concentrate our intellectual and spiritual forces, and negotiate practical real-time solutions to the complex problems we must inevitably confront. We will gradually learn to find our way within a mutating, wandering cosmos, to become, to the extent possible, its authors, to collectively invent ourselves as a species. Collective intelligence is less concerned with the self-control of human communities than with a fundamental *letting-go* that is capable of altering our very notion of identity and the mechanisms of domination and conflict, lifting restrictions on heretofore banned communications, and effecting the mutual liberation of isolated thoughts.

We are thus in the situation of a species whose members have good memories, are observant and clever, but who have not yet achieved the state of collective intelligence of the culture for lack of an articulated language. How do you invent a language that no one has spoken, for which there are no records, no examples, and when we lack even an idea of what such a language might be? We can apply the analogy to our present situation: We don't know what we are supposed to create, although we may have already begun to sketch its outline obscurely. Over the course of several millennia, however, *Homo habilis* became *sapiens*, crossed a similar threshold, went forth into the unknown, invented the earth, the gods, and the infinite world of signification.

But languages are made for communication within small communities "on a human scale" and perhaps to enable the development of relations between such groups. Through writing we have entered a new stage of our evolution. This technology has led to the increased efficiency of communication and the organization of human groups much larger than ordinary speech could have accommodated. There was a price, however, namely the division of society into a bureaucratic information processing machine based on writing on the one hand and "administered" individuals on the other. The problem faced by collective intelligence is that of discovering or inventing something beyond writing, beyond language, so that the processing of information can be universally distributed and coordinated, no longer the privilege of separate social organisms but naturally integrated into all human activities, our common property.

This new dimension of communication should obviously enable us to share our knowledge and acknowledge it to others, which is the fundamental condition for collective intelligence. Beyond this are two major possibilities, which could radically transform the fundamental data of social life. First, we will have at our disposal simple and practical means for knowing what we are doing as a group. Second, we will be able to manipulate, much more easily than we are able to write, the instruments for collective utterance. This will no longer take place on the scale of some Paleolithic clan, or that of the state and the historical institutions of the nation, but in keeping with the size and speed of the enormous turbulence, deterritorialized processes, and anthropological nomadism that we are now subject to. If our societies are content merely to be intelligently governed, it is almost certain that they will fail to meet their objectives. To have a chance for a better life, we must become collectively intelligent. Transcending the media, airborne machines will announce the voice of the many. Still indiscernible, cloaked in the mists of the future, bathing another humanity in its murmuring, we have a rendezvous with the *over-language*.

Collective Intelligence

Mankind's Emerging
World in Cyberspace

Introduction

Economy

The prosperity of a nation, geographical region, business, or individual depends on their ability to navigate the knowledge space. Power is now conferred through the optimal management of knowledge, whether it involves technology, science, communication, or our "ethical" relationship with the other. The more we are able to form intelligent communities, as open-minded, cognitive subjects capable of initiative, imagination, and rapid response, the more we will be able to ensure our success in a highly competitive environment. Our material relationship to the world is maintained through a formidable epistemological and logical infrastructure: educational and training institutions, communications networks, digitally supported intellectual technologies, the continuous improvement and distribution of skills. In the long term, everything is based on the flexibility and vitality of our networks of knowledge production, transaction, and exchange.

It would be a gross oversimplification to compare the transition to the age of knowledge with the shift to a service economy. Such a transition cannot be reduced to the displacement of industrial activities to the service sector, for the service sector itself is increasingly coming under siege by a variety of technological objects. It is becoming "industrialized," as characterized by the presence of ATMs, Web sites, educational software, expert systems, etc. To a greater and greater extent, industrial organizations see their activities as a form of service. To respond to the new conditions of economic life, businesses tend to organize themselves in such a way that they are receptive to *innovation networks*. This means, for example, that in a large corporation, departments can easily and quickly interact with one another, without the need for any kind of formal agreement, and with the continuous exchange of information and personnel. Interactive systems and contemporary innovation networks intersect one another, operating across the enterprise. The increasing growth of partnerships and alliances is a striking illustration of this process. New abilities must continuously be imported, produced, and introduced (in real time) in all sectors of the economy. Organizations must remain receptive to a constantly renewed stream of scientific, technical, social, and even aesthetic skills. Skill flow conditions cash flow. Once the process of renewal slows down, the company or organization is in danger of petrifaction and extinction. As Michel Serres has written, knowledge has become the new infrastructure.

Why did the so-called communist governments begin to decline sharply during the seventies, before finally collapsing at the beginning of the nineties? Without going into too many details on what is a complex issue, I can offer one hypothesis[1] that may be able to shed considerable light on our approach to the age of knowledge. The bureaucratically planned economy, which was still capable of functioning at the beginning of the sixties, was incapable of following the transformation of labor that resulted from the contemporary evolution of technological and organizational structure. Totalitarianism collapsed in the face of new forms of mobile and cooperative labor. *It was incapable of collective intelligence.*

The great shake-up of Western economies toward the tertiary sector was not the only factor involved in this, however. A more significant movement was under way, one that was anthropological. Beginning in the nineteen sixties, it became increasingly difficult for a laborer, employee, or engineer to inherit the traditions of a trade, to exercise and transmit this ability almost unchanged, to assume a lasting professional identity. Not only did technologies change with increasing velocity, but it became necessary to learn how to compare, regulate, communicate, and reorganize one's activity. It became necessary to exercise one's intellectual potential on a continuous basis. Moreover, new conditions of economic life gave a competitive edge to organizations in which each member was capable of taking the initiative for coordination, rather than submitting to some form of top-down planning. But this constant mobilization of social and cognitive abilities implicitly assumed a considerable degree of subjective involvement. No longer was it sufficient to passively identify oneself with a category, trade, or community. Now one's uniqueness, one's personal identity were implicated in professional life. It is precisely this form of subjective mobilization, highly individual as well as ethical and cooperative, that the bureaucratic and totalitarian universe was incapable of generating.

Quite obviously the interpenetration of leisure, culture, and work as a form of global social and subjective commitment remains the privilege of business leaders, the more highly qualified executives, certain professions, researchers, and artists. There are indications, however, that this model will expand, "trickle down" by a process of capillary motion, to all layers of society. The fact that the boundary between our professional life and personal development is beginning to blur signifies the death of a form of economic activity. Economic goals and technological efficiency can no longer operate within a closed circuit. As soon as genuine subjective commitment is required of individuals, economic needs must give way to politics in the broadest sense of the word, that is to ethics and civic responsibility. They must also reflect cultural significations. Pure economy and mere efficiency cease to become effective. Only by incorporating cultural and moral objectives, aesthetic experience, can business engage the subjectivity of its employees, as well as its customers. The corporation no longer only consumes and produces goods and services, as in traditional economics. It is no longer satisfied with implementing, developing, and distributing skill and knowledge, as illustrated by the new cognitive approach to organizational structure. We must recognize the fact that the corporation, like other institutions, both encourages and promotes the development of subjectivity. Because it conditions all other activities, the continuous production of subjectivity will most likely be considered the major economic activity throughout the next century (see Chapter 2).

Under the wage system the individual sells his physical strength or labor time within a quantitative and easily measurable framework. Such a system could easily give way to a form of self-promotion, involving qualitatively differentiated abilities, by independent producers or small teams.[2] Individuals and microcorporations are more capable than large companies of continuous reorganization and optimal enhancement of the individual skills that are currently the requirements for success. Economic life will no longer be driven primarily by competition among large companies, which encourage quantitative and anonymous forms of labor. Rather, we are witnessing the development of complex forms of confrontational interdependence among skill zones that are fluid, delocalized, based on their singularities, and agitated by permanent molecular movements of association, exchange, and rivalry. The ability to rapidly form and reform intelligent communities will become the decisive weapon of regional skill centers competing within a globalized economic space. The emergence and constant redefinition of distributed identities will not only take place within the institutional framework of business, but through cooperative interactions in an international cyberspace.

Anthropology

Once knowledge becomes the prime mover, an unknown social landscape unfolds before our eyes in which the rules of social interaction and the identities of the players are redefined. A new anthropological space, the *knowledge space*, is being formed today, which could easily take precedence over the spaces of earth, territory, and commerce that preceded it. The second part of this book (Chapters 7–15) is devoted to a detailed cartography of these spaces and their interrelationship.

What is an anthropological space? It is a system of proximity (space) unique to the world of humanity (anthropological), and thus dependent on human technologies, significations, language, culture, conventions, representations, and emotions. For example, in the anthropological space I refer to as "territorial," two individuals, living on either side of a border, are "farther" from one another than from people living in the same country, while this relationship might be reversed in the space of physical geography.

The *earth* was the first great space of signification formed by our species. It is based on the three primordial characteristics that distinguish *Homo sapiens*: language, technology, and complex forms of social organization ("religion" in the broadest sense of the word). Only humanity lives on *this* earth; animals inhabit ecological niches. Our relationship to the cosmos is the fundamental aspect of this first space, both from a point of view that we would today qualify as imaginary (animism, totemism), as well as from a very practical point of view, given the intimate contact between us and "nature." Myth and rite are the specific modes of knowledge of this first anthropological space. On earth, identity is inscribed within our bond to the cosmos as well as in our affiliation or alliance with other men. The first item on our resumé is generally our name, our symbolic inscription within an ancestral line.

A second, *territorial* space arose during the Neolithic period with the development of agriculture, the city, the state, and writing. This second space did not eliminate the great nomadic earth but partially covered it and attempted to turn it into something sedentary, domesticated. Hunting and gathering were no longer a source of wealth, but the possession and exploitation of fields. Within this second anthropological space the dominant modes of knowledge were based on writing: history and the development of systematic, theoretical, and hermeneutic knowledge. Here, the pivot of existence was no longer participation in the cosmos but the link to a territorial entity (affiliation, property, etc.) defined by its borders. Today, along with our name, we have an address, which serves to identify us within the territory of residents and taxpayers. The institutions in which we live are also territories, or juxtapositions of territories, with their hierarchies, bureaucracies, systems of rules, borders, logic, belonging, and exclusion.

A third anthropological space began to develop in the sixteenth century, which I will call the *commodity* space. It began to take shape with the initial development of a world market following the conquest of America by Europeans. The organizing principle of the new space is movement: the flow of energy, raw materials, merchandise, capital, labor, information. The great movement of *deterritorialization* that began to develop at the dawn of the modern era did not result in the suppression of territories but in their subversion, their subordination to economic flux. The commodity space did not eliminate the preceding spaces, but outpaced them. It became the new engine of evolution. Wealth was no longer based on controlling borders but on the control of movement. Industry now rules, in the general sense of processing materials and information. Modern experimental science is a typical mode of knowledge of the new space of continuous movement. But traditional science is itself undergoing a process of deterritorialization. Following the Second World War, it gave way to a "technoscience" driven by a permanent dynamic of research and economic innovation. The coupling of theory and experimental practice characteristic of classical science now had to compete with the growing power of simulation and digital modeling, which threatened conventional epistemological methods and provided a glimpse into the turmoil of a fourth space. To possess an identity, to exist in the space of commodity flow, means that we participate in economic production and exchange, occupy a position at the nodes of the various networks of production, transaction, and communication. To be unemployed within the commodity space is a sign of misfortune, for within it our social identity is defined by work, which means, for the majority of the population, a job and a salary. On our resumé, right after our name (position on earth) and address (position within the territory), we generally indicate our profession (position in the commodity space).

Is it possible to bring a new space into existence, in which we would possess a social identity even without a profession? Perhaps the current crisis of identity and social forms of identification signifies the dimly perceived and incomplete emergence of a new anthropological space, that of knowledge and collective intelligence, whose arrival is in no way guaranteed by any historical laws. Like the other anthropological spaces, the *knowledge space* will control preceding spaces rather than eliminate them. From this point forward, the existence of economic networks and territorial power will depend on mankind's capacity for the rapid acquisition of knowledge and the development of a collective imagination, as will the survival of the great nomadic earth.

Intelligence and human knowledge have always played a central role in social life. Our species is called *sapiens* for good reason. To each anthropological space there corresponds a specific mode of knowledge. But then, why refer to civilization's new horizon as the *knowledge space*? There are at least three aspects to this newness: the rate of evolution of knowledge, the number of people who will be asked to learn and produce new forms of knowledge, and finally, the appearance of new tools (cyberspatial tools) capable of bringing forth, within the cloud of information around us, unknown and distinct landscapes, singular identities characteristic of this space, new sociohistoric figures.

Speed. Never before has science and technology evolved so rapidly, with so many direct consequences on our daily life, work, modes of communication, our relation to our bodies, space, etc. Today it is within the universe of knowledge and skill that acceleration is greatest and the configurations most changeable. This is one of the reasons why knowledge (in the most general sense of the word) dominates the other dimensions of social life.

Mass.   It has become impossible to restrict knowledge and its movement to castes of specialists. From now on, humanity as a whole must adapt, learn, and invent if it is to improve its lot in the complex and chaotic universe in which we now live.

Tools.   The number of messages in circulation has never been as great as it is now, but we have few instruments to filter the pertinent data, make connections on the basis of significations and needs that are still subjective, or orient ourselves within the flux of information. It is at this point that the knowledge space ceases to be the object of established fact and becomes a project. Building the knowledge space will mean acquiring the institutional, technical, and conceptual instruments needed to make information navigable, so that each of us is able to orient ourselves and recognize others on the basis of mutual interests, abilities, projects, means, and identities within this new space. The deliberate creation of a system of expression for the knowledge space will enable us to correctly express, and perhaps even resolve, a number of crucial problems that we are currently unable to formulate adequately with the concepts and tools that have been used to express preceding spaces.

Our living knowledge, skills, and abilities are in the process of being recognized as the primary source of all other wealth. What then will our new communication tools be used for? The most socially useful goal will no doubt be to supply ourselves with the instruments for sharing our mental abilities in the construction of collective intellect or imagination. Internetworked data would then provide the technical infrastructure for the collective brain or *hypercortex*[3] of living communities. The role of information technology and digital communications is not to "replace mankind" but to promote the construction

of intelligent communities in which our social and cognitive potential can be mutually developed and enhanced.
Based on this approach, the major architectural project of
the twenty-first century will be to imagine, build, and
enhance an interactive and ever changing cyberspace (see
Chapter 6). Perhaps it will then be possible to move beyond the society of the spectacle and enter a post-media
era in which communications technologies will serve to
filter and help us navigate knowledge, and enable us to
think collectively rather than simply haul masses of information around with us. Unfortunately, although the pro-
moters of the information highway may be aware of the
problem, they remain mired in discussions about band-
width. Fortunately, at present only a small minority con-
siders the global system for delivering video on demand
to be the *nec plus ultra* of imaginative thought concerning
the art and architecture of cyberspace.

The Social Bond and
Its Relationship to Knowledge

In addition to the required technical instrumentation,
the project for a knowledge space will lead to a re-creation
of the social bond based on reciprocal apprenticeship,
shared skills, imagination, and collective intelligence. It
should be obvious that collective intelligence is not a
purely cognitive object. Intelligence must be understood
here in its etymological sense of joining together (*inter
legere*), as uniting not only ideas but people, "constructing
society." It covers a very general approach to social life
and its possible future. Collective intelligence, as the term
is used in this book, is a global project whose ethical and
aesthetic dimensions are as important as its technological
and organizational aspects. This ethical approach will be
more fully developed in Chapters 1 and 5. Given the sin-

gular lack of perspective characteristic of our own epoch, I would like to risk the suggestion that we follow a new bearing, a new direction, a kind of utopia. This vision of the future is organized around two complementary axes: the renewal of the social bond through our relation to knowledge and collective intelligence itself.

The question of the construction or reconstruction of the social bond is especially difficult at a time when elements of humanity seem to be imploding and metastasizing and in the process of losing any fixed frame of reference and sense of identity. It is essential that we explore other methods of social integration aside from increasingly scarce forms of salaried employment. The need to develop alternatives is made more urgent given the fact that the production of communities through ethnic, national, or religious adhesion has led to deadly confrontations that have become all too common. By establishing the social bond on the basis of our relationship to knowledge, we will encourage the growth of a *deterritorialized civility* that coincides with contemporary sources of power while incorporating the most intimate forms of subjectivity.

Through our interactions with things, we develop skills. Through our relation to signs and information, we acquire knowledge. Through our relationship to others, mediated by processes of initiation and transmission, we bring knowledge to life. Skill, understanding, and knowledge (which can all refer to the same objects) are three complementary modes of cognitive transaction and continuously interpenetrate one another. Each activity, each act of communication, each human relation implies an apprenticeship. By means of the skills and understanding that it envelops, a life can continuously feed a circuit of exchange, nourish a sociability of knowledge.

We can explicitly establish, openly and publicly, mutual apprenticeship as a way of mediating relationships among individuals.[4] Our identities would then become knowledge identities. The ethical consequences of this new institution of subjectivity are immense. Who is the other? Someone who has knowledge. And moreover, someone who knows what I do not. The other ceases to be the terrifying, threatening figure he is now; like me, he is ignorant of much and possesses limited knowledge. But since our zones of inexperience do not overlap, he represents a source of possible enrichment for my own understanding. He can augment my powers of being, and this will increase to the extent that we are different. I could combine my skills with his in such a way that we could work better together than apart. The "knowledge trees"[5] that are used in businesses, schools, and other organizations, enable us to encounter the other as a bundle of knowledge within the knowledge space and no longer as a name, address, profession, or social status.[6]

But the transparency will never be total, nor should it be. Knowledge of the other can't be reduced to a sum of results or data. Knowledge, in the sense we are using the word here, is also a knowledge-of-living; it is inseparable from the construction and habitation of a world, and incorporates the full span of our life. For this reason, even though I need to gather information and exchange ideas, even if I am able to learn from the other, I'll never know everything he knows. Our need to listen to the other can never lead to the construction of knowledge about him. We cannot simply capture his expertise or the information he possesses. Apprenticeship, in the fullest sense of the word, also implies that we confront the incomprehensibility, the irreducibility of the world of the other, which is the basis of my respect for him. Although a possible source of power for me, the other remains enigmatic, becomes a desirable being in every respect.

If the other is a source of knowledge, reciprocity is immediate. Regardless of my temporary social position, regardless of the judgment of an educational institution about my abilities, I can also become an opportunity for learning to someone else. Through my experience of life, my professional career, my social and cultural habits, I can—since knowledge is coextensive with life—provide knowledge resources to the community. Even though I am unemployed, or without money or a diploma, condemned to life in a ghetto, illiterate, I am not useless. I am not interchangeable. I have an image, a position, dignity, a personal and positive value within the knowledge space. All of us have the right to be acknowledged as a knowledge identity.

The knowledge space is brought to life whenever we initiate human relations based on ethical principles. These include individual improvement through skills acquisition, the efficient transformation of difference into collective wealth, the integration of the exchange of knowledge within a dynamic social process in which each of us is recognized as a unique individual and is not prevented from learning by programs, prerequisites, *a priori* classifications, or prejudices about what is and is not worthwhile knowledge.

What Is Collective Intelligence?

What is collective intelligence? It is a form of *universally distributed intelligence*, constantly enhanced, coordinated in real time, and resulting in the effective mobilization of skills. I'll add the following indispensable characteristic to this definition: The basis and goal of collective intelligence is the mutual recognition and enrichment of individuals rather than the cult of fetishized or hypostatized communities.

My initial premise is based on the notion of a universally distributed intelligence. No one knows everything, everyone knows something, all knowledge resides in humanity. There is no transcendent store of knowledge and knowledge is simply the sum of what we know. The light of the mind shines even where we attempt to persuade others that no intelligence exists: "educational failure," "rote execution," "underdevelopment." The overarching judgment of ignorance turns against the judges. If you are tempted to judge someone as ignorant, look for the context in which his knowledge can be turned into gold.

Intelligence is constantly enhanced. There is little doubt that intelligence is universally distributed. But facts must find some form of concrete realization. An intelligence that is frequently ridiculed, ignored, unused, and humiliated is obviously not enhanced. While we are increasingly concerned with economic and ecological waste, it seems as if we are willing to squander our most precious resource by refusing to acknowledge it, develop it, or even use it when it is found. From a school report card to a corporate job profile, from archaic management methods to social exclusion through unemployment, we are currently witnessing the deliberate organization of ignorance concerning the extent of the intelligence around us, a terrifying waste of experience, skill, and human wealth.

The real-time coordination of intelligence. This will involve communications methods that, once past a certain quantitative threshold, must be based on digital information technologies. New communications systems should provide members of a community with the means to coordinate their interactions within the same virtual universe of knowledge. This is not simply a matter of modeling the conventional physical environment, but of enabling members of delocalized communities to interact within a mobile landscape of signification. Events, decisions, actions, and individuals would be situated along dynamic maps of shared context and continuously transform the virtual universe in which they assume meaning. In this sense cyberspace would become the shifting space of interaction among knowledge and knowers in deterritorialized intelligent communities.

The effective mobilization of skills.   Before we can mobilize skills, we have to identify them. And to do so, we have to recognize them in all their diversity. Officially validated skills now represent only a tiny minority of those that are active. The question of recognition is critical. Not only does it lead to improved skills administration in business and community environments, it also possesses an ethical and political dimension. In the age of knowledge, failure to recognize the other as an intelligent being is to deny him a true social identity. It feeds resentment and hostility, the humiliation and frustration from which violence is born. For when we acknowledge the other for the range of skills he possesses, we allow him to identify himself in terms of a new and positive mode of being, we help mobilize and develop feelings of recognition that will facilitate the subjective implication of other individuals in collective projects.

The ideal of collective intelligence implies the technical, economic, legal, and human enhancement of a universally distributed intelligence that will unleash a positive dynamic of recognition and skills mobilization. One of the necessary conditions for the economic rise of Europe at the end of the eighteenth century was the introduction of an effective legal guarantee for intellectual property (copyrights, licenses, patents of invention, etc.). Thus inventors could devote their time, intellectual energy, and financial resources to innovation without worrying about being stripped of the results of their efforts by those in power. Once monopoly and economic privilege were banished by law, as soon as there were methods to assign, publicly and irreversibly, the mark of a physical or moral person to a technical process, innovation became worthwhile for its own sake. Once rules were in place for the process of innovation, once it became a legitimate activity, one that was socially encouraged and economically recompensed, a scientific and industrial dynamic of immense scope was set in motion. Yet we are now faced with the necessity of making a similar transition with respect to the skills and intelligence of collectivities, for which no system of measurement currently exists, no method of accounting, representation, or legal regulation worthy of the name, although they are the source of all contemporary forms of power.

It is worth bearing in mind that collective intelligence is a universally distributed intelligence that is enhanced, coordinated, and mobilized in real time. To prevent any possible misunderstanding, I would also like to specify what it isn't. Collective intelligence must not be confused with totalitarian projects involving the subordination of individuals to transcendent and fetishistic communities. In an ant colony, the individuals are "dumb;" they have no collective vision and no awareness of how their actions are integrated with those of other individuals. But although individual ants might be "stupid," their interaction results in an emergent behavior that is globally intelligent. Yet the ant colony possesses a rigidly fixed structure; the ants are sharply divided into castes and are interchangeable within those castes. The ant colony is the opposite of collective intelligence in the sense that I am using the expression. Far from leading us in the direction of the knowledge space, the ant colony precedes the earth, it is prehuman. Any attempt to assimilate the operation of society to that of an ant colony should be considered barbarous and reprehensible.

Collective intelligence is born with a culture and grows with it. Obviously, when we think, we make use of ideas, languages, and cognitive technologies inherited from a community. But a culturally informed intelligence is no longer programmed like that of a termite colony or bee hive. Through processes of transmission, invention, or forgetfulness, heritage becomes an element of individual responsibility. The intelligence of the group is no longer the mechanical result of blind or automatic activities, for it is individual thought that perpetuates, invents, and mobilizes that of society. And yet, the intelligent community described in this book cannot be reduced to the status of conventional culture. In an intelligent community the specific objective is to permanently negotiate the order of things, language, the role of the individual, the identification and definition of objects, the reinterpretation of memory. Nothing is fixed. Yet, this does not result in a state of disorder or absolute relativism, for individual acts are coordinated and evaluated in real time, according to a large number of criteria that are themselves constantly reevaluated in context.[7] In place of the "invisible hands" of the termite colony, we have the visible hands and imaginable dynamic of expanding virtual universes. Through their interaction with diverse communities, the individuals who animate the knowledge space are, far from being interchangeable members of immutable castes, singular, multiple, nomadic individuals undergoing a process of permanent metamorphosis (or apprenticeship).

This project implies a new humanism that incorporates and enlarges the scope of self knowledge into a form of group knowledge and collective thought. The old adage "I think, therefore, I am" is generalized as a process of collective intelligence leading to the creation of a distinct sense of community. We pass from the Cartesian *cogito* to *cogitamus*. Far from merging individual intelligence into some indistinguishable magma, collective intelligence is a process of growth, differentiation, and the mutual revival of singularities. The shifting image that emerges from such skills and projects, and from the relations among members in the knowledge space, constitutes, for a community, a new mode of identification, one that is open, dynamic, and positive. New forms of democracy, better suited to the complexity of contemporary problems than conventional forms of representation, could then come into being (see Chapter 4).

The first part of this book is devoted to the process of engineering the social bond, which is the process of creating intelligent communities and maximally enhancing the diversity of human abilities. Collective intelligence is described from several points of view: ethical (Chapters 1 and 5), economic (Chapter 2), technological (Chapter 3), political (Chapter 4), and aesthetic (Chapters 5 and 6). The heart of the process is the economy of human abilities. That the power of messages, machinery, and natural kinds might finally be evaluated, exploited, and accounted for in terms of this subjective economy, that the value of things might be expressed in terms of the same signs used for individual identity (rather than the other way around), that every aspect of our environment might become human, this is the utopia within the utopia traced by the process of engineering the social bond.

Part Two of this book, "The Knowledge Space," develops the theory of the four anthropological spaces mentioned in the introduction. After presenting the earth, the territory, the commodity space, and the knowledge space (Chapter 7), the concept of an anthropological space is defined (Chapter 8) before a review of the problems of identity (Chapter 9), signification (Chapter 10), space and time (Chapter 11), followed by a more detailed discussion of the question of knowledge (Chapters 12, 13, and 14). Part Two ends with an outline of political philosophy conceived as a theory of the relations among the anthropological spaces (Chapter 15). While the linearity of text occasionally requires us to present ideas in chronological order, "The Knowledge Space" can be seen as a type of cartography, a conceptual toolbox, a portable guide to anthropological mutation rather than a history. This text will hopefully serve as a kind of manual for the changes now under way, and as a means to locate obstacles and indicate potential directions for exploration. I do not claim historical or scientific accuracy, but philosophical and practical fecundity.

I

Engineering the Social Bond

1

The Just

The Ethics of Collective Intelligence

Genesis, chapters 18 and 19. A great cry rises up against Sodom and Gomorrah. Having determined to destroy the cities, the source of so much injustice, God first decides to speak to Abraham. Though he is nothing but dust and ash before God, the patriarch's dialogue with the Eternal turns into an extraordinary bargaining session. "If there are fifty good men in the city, will you destroy Sodom, will you cause the good to perish with the bad?" God grants Abraham the salvation of the city, providing he can find fifty good men. But the patriarch remains stubborn and continues to negotiate for the salvation of the city. From fifty to forty-five, then thirty, then twenty, and finally only ten good men.

At nightfall two angels arrive at the gates of Sodom. Nothing about their appearance indicates that they are the envoys of God. To the townspeople they are unknown travelers, just passing through. Lot, who had been sitting near the entrance to the city, invites the strangers to his home, feeds them, and treats them according to the laws of hospitality. They hadn't yet gone to bed when the people of Sodom gathered around Lot's house, demanding the strangers that they might "know them." Lot refuses to turn over his guests, even going so far as to offer his daughters in exchange to the angry crowd. But the crowd remained insistent. This test enabled God to determine the number of good men in Sodom: there was but one. The angels organized the escape of Lot and his family. As soon as they had left, the city was destroyed. Despite the angels' warning, Lot's wife turned back to look at the rain of fire and brimstone that fell upon Sodom and Gomorrah. She was immediately changed into a pillar of salt.

I would like to venture a lay interpretation of the biblical tale of the destruction of Sodom and Gomorrah. The text is, in fact, open to such an interpretation. What it presents is not so much a transcendent principle of good and evil as the power of living individuals, the "just," who are capable of maintaining the existence of the human world.

If we consider Lot's wife as his other "half," her fate illustrates the temptation of the just to *linger at the point of judgment* rather than welcome the human other. Through his wife, Lot identifies with the judge, or even with the abstract principle of justice, rather than a living but just man. Lot's wife turns back toward the inferno in which the inhabitants of the cities are dying and, having done so, reifies practice as a transcendent value. The just maintain life; the judges turn to stone. At any moment the just individual might forget himself and be transformed into a pillar of salt, as stiff and motionless as justice.

We can assume that the barter between God and Abraham takes place throughout time and space. If our

human world has lasted until the present, it is because the
just are still with us, because the traditions of hospitality,
assistance, openness, care, recognition, and constructive-
ness are, in the end, more common or stronger than exclusion, indifference, negligence, resentment, and destruction. If parents failed in their responsibility to their
children, if people spent their time in jealousy, rancor,
and killing, then the human species would have ceased
to survive. In reality the rain of fire and brimstone that
burned Sodom and Gomorrah didn't fall from the heavens;
it rose from the cities themselves. These were the flames
of discord, war, and violence that swirled around the
inhabitants. But not all cities were destroyed, and our
presence on earth proves that, until now and throughout
the world, the "quantity of good" has exceeded the
"quantity of evil." This assessment in no way justifies the
suffering and degradation humanity undergoes in obtaining some final good. It is simply intended to balance
the widespread recognition of evil by the consideration of
a fact, a piece of raw data: We continue to exist. The
human megalopolis has not yet been destroyed.

Evil is everywhere and always visible, while the
good (the activity of the just) can only be discovered after
a detailed, on-the-spot investigation (the angels visit
Sodom), or by its indirect effects, the culmination of a
painstaking process of reasoning. The biblical text is clear
on this point: God hears the uproar, the cries, the pleas
that rise up against Sodom and Gomorrah. He is initially
warned of their injustice. The voice of criticism is the first
to be heard. Civil war, assassination, dictatorship, misfor-
tunes of all sorts are the substance of television programs
and are splashed across the headlines of our newspapers.
God is kept up to date about evil. Yet, when Abraham
began to negotiate the number of the just needed to save
the city, we realize that even He didn't know whether

there were fifty, or forty-five, or thirty, or twenty, or ten, or fewer. God's science (that is, according to a lay interpretation, human science) does not extend that far. While evil is outspoken, the number and even the identity of the just are unknown, hypothetical. Evil is newsworthy, but the just are concealed, discreet, anonymous, ignored. How then do we recognize the just? In the text there is no great tribunal, no final judgment, no weighing of souls. There are only a couple of migrants, who travel the world and arrive one evening, covered with dust from their trip, at the gates of a city. We recognize the just by moving from place to place. There is no transcendent justice or omniscience that will enable us to find them. We must follow the nomads. They are in search of the invisible beings who support the world. They can identify the just who weave the social bond in the depth of a shadow.

What was Sodom's crime? The refusal of hospitality. Rather than welcome the strangers, the Sodomites wanted to rape them. Hospitality is the perfect representation of the maintenance of the social bond, one conceived in accordance with the formula of reciprocity: we are likely to be either the receiver or the received. And each of us may become a stranger in turn. Hospitality sustains the possibility of travel, of meeting the other. Through hospitality, he who is lost, different, foreign, is welcomed, integrated, included in a community. Hospitality is the act of attaching the individual to a community. In every respect, it is the opposite of exclusion. The just man includes, he integrates, he repairs the social fabric. In a society of the just, and in accordance with the conventions of reciprocity, everyone strives to include the other. In a world where everything is in flux, in which everyone is faced with change, hospitality, the morality of nomads and migrants, becomes the very essence of morality. Yet the fact that he works to weave together the nomadic community does not mean that the just individual is willing to pay any price to achieve unity, uniformity, or unanimity. Lot, for example, took the risk of being in the minority, the smallest minority possible since he was alone in defending the strangers against the others. He put himself in the position of a stranger. The most fully inclusive can become the most excluded. Helping to integrate the foreigner, being hunted in turn, ferrying others across borders, the just individual is the quintessential smuggler.

Why didn't Abraham continue to barter (nine good men, seven, three …)? Why were ten people needed for the city to be spared? Why didn't Lot succeed in saving Sodom? Because collective force is required to sustain a community. For there to have been three, this would have implied that three people were known, and soon celebrated, but sooner or later *one* of them would end up being separated from the rest. But this text, far from being a representation or spectacle, demonstrates that in reality a city cannot be founded on the relation of the group to the individual or the individual to the group. It is not the function of the just individual either to rule or to serve as scapegoat. The city is based on the relation of one community to another. Ideally, it survives through its relationship with itself, through the work of material inclusion practiced by all. Yet ten people are the beginning of a real community. With ten people anonymity is possible. At least ten individuals are required because the just must first pass the test of *the society of the just*. They must be capable of living together, of mutually stimulating and enhancing their acts. The just are efficient; they are able to maintain the existence of a community, only when they form a collective intelligence.

Until now we have been looking at the ways in which the just are able to create and sustain human communities. By showing that the effectiveness of the just individual is to maintain the existence of a community or delay its destruction, the biblical text provides us with critical insight into the nature of the good in general. The good engenders and enhances human qualities. The forces that lead to the creation and preservation of social life are good in all their variety. If the just are able to prevent destruction, it is because the good is closely associated with being, and more importantly, the capacity for being, or strength. And perhaps it is still more closely associated with the increase of strength, whether it be physical, moral, intellectual, sensual, or some other form. Anything, therefore, that causes the growth of human beings would be judged good, and primarily moral: dignity, recognition, communication, collective intelligence. The just promote strength. Conversely, those forces that diminish and eventually destroy humanity will be judged as bad: humiliation, depreciation, separation, isolation. If strength is good, *power* is bad, for it is measured by its capacity to limit strength, by its potential for destruction. Power creates fear. Power is boisterous; it prevents the community from communicating with itself. Power comes into being and sustains itself only by impoverishing the qualities of being around it. The just avoid power.

Forever in the presence of being and strength, the just compete to produce and maintain all that populates the human world. Through them, although their names are never mentioned, the world is able to function; its substance is created and preserved: doting mothers, ghost writers, maids, secretaries, workers who keep the factory going in spite of the engineers' blueprints, and all those who repair the machinery, reunite households, break the chains of scandal, smile, praise, and listen, enable us to live intelligently. Abraham is the epitome of the just man. He is not content to do good as an individual, he strives to give the greatest scope to the acts accomplished by other just individuals. By convincing God that ten good men could save the city, he enhances and maximizes the potential for good; he turns our attention to the good of others. Abraham's bargaining with God is the earliest technology for optimizing activity, for the maximal exploitation of the minimum number of positive qualities found in a human community. Abraham invented the process of engineering the social bond.

2

Human Qualities

The Economy of Collective Intelligence

The information society is a trap. The economy, we are told, after being based on agriculture, then industry (the transformation of matter), is now being driven by information processing. But as numbers of employees and managers have found out, nothing can be automated as thoroughly and as quickly as the processing and transmission of information. What remains after we have mechanized agriculture, industry, and messaging technologies? The economy will center, as it does already, on that which can never be fully automated, on that which is irreducible: the production of the social bond, the relational. This implies not only an economy of knowledge but a more general, human economy, one that comprises the economy of knowledge as one of its subsystems.

In principle the activities involved in the production of goods and services should have as their goal the enrichment of humanity, an increase of our strength in the sense given above. For example, they can increase individual and group skills, promote sociability and mutual awareness, supply the tools for autonomy, create diversity, and vary the kinds of leisure activities we engage in. Yet, anything that can be thought of as only a moral imperative and, therefore, optional with respect to the goals of traditional economy, tends to become a limiting obligation, a condition of success. The continuous transformation of technology, markets, and the economic environment leads communities to abandon rigid and hierarchical methods of organization and develop the capacity for initiative and active cooperation among their members. Accomplishing this, however, requires the effective mobilization of the subjectivity of those individuals. Whether at the corporate or administrative level, whether it be regional or national in scope, the tension toward collective intelligence assumes that we are willing to focus on the human as an end in itself. When we begin to draw on the emotional and intellectual resources of individuals and cultivate our capacity for fully recognizing the other, when global relations and social interactions create a situation in which competition is based on common interest, then competition shifts to the domain of ethics.

Bureaucratic and totalitarian communities, societies that are diseased with corruption and crime, weaken the foundations on which the new conditions for economic success are established. Given equal material resources and similar economic constraints, victory will be claimed by those groups whose members work for their enjoyment, learn quickly, live up to their commitments, respect themselves and others, and move freely throughout a territory rather than trying to control it. Those who are the most just, the most capable of fashioning a collective intelligence together will succeed. Thus effectively and subjectively lived human experience is not only the theoretical end of economic activity, but becomes the very condition for its existence. Economic necessity is conflated with the need for ethics. Yet there are other reasons to encourage the development of an economy of human qualities.

The growth of technology, the advance of science, geopolitical turbulence, and the randomness of markets are dissolving traditional trades and destroying communities, giving rise to regional revolutions, forcing people to abandon their homes, their country, their customs, and even their language. Deterritorialization is often the fabric of exclusion; it destroys social bonds. In almost all cases it blurs identities, at least those based on a sense of adhesion, on "roots." The result is complete disarray, and an immense need for community, for bonds, for recognition and identity. Such conditions are fertile ground for racism, fundamentalism, nationalism, and crime. In contrast, and just as nuclear energy gave rise to an industry for recycling radioactive waste, accelerated deterritorialization will spawn an industry for reconstructing the social bond, the rehabilitation of the outcast, the recasting of identities for individuals and communities without structure. It is not only for reasons of economic competition, but also out of a genuine sense of social urgency that development of the sector responsible for producing the social bond (one of the principal activities in the economy of human qualities) is fostered. And the conditions that favor the new economy will last for quite some time.

But there is another reason that makes it necessary to promote an economy of human qualities and develop the formation of the social bond that accompanies it. In the modern world available technology enables us to provide everyone with more than the bare necessities of life. We are forced to conclude that scarcity is socially manufactured, that poverty and exclusion are organized, even if they are not intentionally promoted. While unemployment can be viewed as a more or less understandable, though unfortunate occurrence according to traditional economic theory, in terms of human economy, it represents the systematic destruction of wealth. A society that explicitly acknowledges the principles of an economy of human qualities will recognize, encourage, and maintain those qualities, even those that do not play a direct role in the system of commodity production. In this way it would enable those without employment to construct an identity for themselves through interaction with the community. Moreover, it would indirectly enrich our store of skill and human potential, which promotes the dynamism of the commodity sector.

Yet neither the economy of knowledge nor the enlarged economy of human qualities should be allowed to develop as directed economies, for this would imply the use of means that are radically opposed to our stated objectives. Non-commodity does not necessarily signify something that is state-run, bureaucratic, monopolistic, hostile to private initiative, or foreign to any form of evaluation. The problem of engineering the social bond requires that we invent and maintain methods for regulating a generalized liberalism. In keeping with this expanded notion of liberalism, each of us would be an individual producer (and consumer) of human qualities in a wide variety of markets or contexts, while no one would ever be able to appropriate the means of production exclusively for their own use. In the economy of the future, capital will be the total individual.

Those who manufacture things will become scarcer and scarcer, and their labor will become mechanized, augmented, automated to a greater and greater extent. Information processing skills will no longer be needed, for intelligent networks will soon be able to function with little human assistance. The final frontier will be the human itself, that which can't be automated: the creation of sensible worlds, invention, relation, the continuous re-creation of the community.

In spite of their diversity nearly all contemporary trades are centered on processes of active cooperation, relation, training, and continuous apprenticeship. Do industrial manufacturers produce goods? Obviously. Yet they spend the majority of their time listening to their customers, negotiating with them, educating them, establishing partnerships, developing their own skills, etc. The police are obviously responsible for the prevention and suppression of crime. Yet at times they also take the place of absent fathers, serve as social workers, social and cultural coordinators, psychologists. Do doctors and nurses care for the body? Of course they do. But relational structures are assuming increased importance. We heal more quickly in humanized hospitals in which the sick are individualized. We can treat patients more effectively by educating them about nutrition, hygiene, and the diagnosis of their symptoms—about medical autonomy in general.

The future of anthropic production is based on two indissoluble factors: the culture of human qualities, primarily skills, and the development of a livable society. It is as if the human, throughout its full extent and variety, were to become the new raw material. I am arguing here for the development of collective intelligence as the ultimate finished product. Collective intelligence will become the source and goal of other forms of wealth, open and incomplete, a paradoxical *output* that is internal, qualitative and subjective. Collective intelligence will be the infinite product of the new economy of the human.

Throughout the Neolithic period, which ended in the middle of the twentieth century, farmers, the majority of our population, worked the earth. During the industrial age, which began at the end of the eighteenth century and is now ending, workers transformed raw materials and employees processed information. Yet the wealth of nations now hinges on our capacity for research, innovation, rapid skills acquisition, and the ethical cooperation of our populations. Those who cultivate human intelligence are thus the source of all prosperity. Today, the new proletariat no longer works with signs or things, but on unprocessed masses of humanity. It accompanies populations as they transit the storms of the great mutation. It humanizes the body, the mind, collective behavior. Even in the midst of battle, it blindly, clumsily forges the weapons of autonomy. These are the new stokers of society: obscure individuals who produce the conditions of wealth far from the light of the spectacle, whose labor is simultaneously the most arduous, the most necessary, and the most poorly paid: the cohort of educators, instructors, professors, and trainers of all kinds. Their ranks will be swelled by the numbers of outreach workers, social workers, police officers, and frustrated prison guards, by masses of auxiliary workers: community groups, non-governmental organizations, charities, support groups, personal aides for a wide range of problems, the army of the unknown that sweeps up after the educational dropouts and gathers the victims of deterritorialization. These new proletarians will be confronted with the relational on a massive scale; they will be the front line of the social bond. The just will be charged with the integration of a discarded population. And through increased mobility and the acceleration of flux, everyone will live on the verge of exclusion, potential outcasts in an unstable universe.

The new proletariat will only free itself by uniting, by decategorizing itself, by forming alliances with those whose work is similar to its own (once again, nearly everyone), by bringing to the foreground the activities they have been practicing in shadow, by assuming responsibility—globally, centrally, explicitly—for the production of collective intelligence, by investing in research on engineering the social bond in order to equip, to the fullest extent possible, those who fashion humanity with their bare hands, exposing the thin skin of affect. The day when the new proletariat attains self-awareness, it will dissolve itself as a class, it will bring about a general socialization of education, training, and the production of human qualities. Unfortunately there is a great temptation toward particularization (to defend what we perceive as our category) rather than singularization. It is easier (in the short term) to cling to archaic images and stable identities than to produce dynamic and changing identities. The reflex to establish networks is acquired more quickly than the effort of creating spaces that are receptive to the circulation of the new nomads.

Yet it is precisely during this period of anthropological nomadism, when we are moving between worlds (rather than across a geographical territory), that transmission and integration can no longer be accomplished exclusively through familial lineage or educational institutions. When there are only a limited number of stable skills in the midst of a massive and continuous change in pertinent forms of knowledge, the channeling of transmission—which served a purpose in other eras—can become a brake on further development. Faced with the deterritorialization of the economic, human, and information flux, and the emergence of anthropological nomadism, our response should be to deterritorialize initiation and humanization itself. Doesn't freedom require means that will strengthen autonomy and enhance the potential of individuals rather than habituate them to dependence? For this reason the transmission, education, integration, and redevelopment of the social bond can no longer be seen as separate activities. They must be carried out by and for society as a whole, and potentially from any point within a fluctuating social entity to any other point, without narrowly channeling those activities or relying on the mediation of specialists. Do we possess the means to achieve such a goal? What kind of engineering will best meet the needs of a growing economy of human qualities?

3

From the Molar to the Molecular

The Technology of Collective Intelligence

Our humanity is the most precious thing we have. It is the source of other wealth, the criterion and living bearer of all value. What use would there be to an object that was never experienced, appreciated, or imagined by a member of our species? Humanity is both the necessary condition of the universe and the superfluity that makes it worthwhile. We are the soil of existence and the extreme measure of its wealth: intelligence, emotion, fragile and protective envelopes without which everything would return to a state of nothingness. For this reason we must economize our humanity; it must be cultivated, enhanced, varied, and allowed to multiply rather than be wasted, destroyed, forgotten, or left to die for lack of care and recognition. While it is good to articulate principles, we must also be prepared to forge new instruments— concepts, methods, technologies—that will make our progress toward an economy of the human something palpable, measurable, organizable, in a word, practical.

There is little doubt that such an economy will be innovative. It will point to the future and be closely tied to the changes now under way. But it is also, and has been since the origin of time, the immemorial foundation of other economies, those whose success determines whether a people falls or flourishes. The engineering of the social bond can't be built from nothing. All the artifices of religion, drugs, perfume, music, images, dreams, trances, stories, prayers, rites, pilgrimages, gatherings, temples, theologies, morals—all are instruments in the toolbox of the priest, the earliest engineer of the social bond. The law, with its codes, constitutions, and jurisprudence, traditional economy, with its accounting systems and models, have developed conceptual, hermeneutic, and mathematical techniques for the rational apprehension, measurement, and regulation of relational facts. How can we turn our backs on this immense store of knowledge? Yet neither religion, law, nor traditional economy have explicitly focused on the open and immanent dynamic of collective intelligence, the maximal enhancement of human qualities, the augmentation and diversification of the strength of being. The evolution of contemporary technology, primarily communication technologies, suggests other approaches, which were inconceivable ten or twenty years ago. This will profoundly affect the range of possible solutions to the problems of managing the social bond and maximizing human qualities. There is nothing revolutionary in the idea that we can engineer the social bond or the notion of human qualities themselves (most of the social sciences can be "applied" in one form or another), the originality lies rather in the ends and mechanisms of this engineering.

In contrast to *molar* technologies, which manage objects in bulk, in the mass, blindly, and entropically, *molecular* technology will manage the objects and processes it

THE MAJOR TECHNOLOGICAL EVOLUTIONS

	Archaic technologies	Molar technologies	Molecular technologies
Control of living species	**Natural selection** Absence of finality Geologic time scale Operates on entire populations	**Artificial selection** Finalization Historic time Operates on entire populations	**Genetic engineering** Finalization Real time Operates on a gene-by-gene basis
Control of matter	**Mechanics** Controls the transmission and point of application of forces Assemblies	**Thermodynamics** (heat) Production of energy and modification of the characteristics of matter by heating and mixing	**Nanotechnology** (cold) Control of the transmission and point of application of forces on a microscopic scale Assembly on an atom-by-atom basis
Control of messages	**Somatics** Production by living beings, variation of messages as a function of context	**Mediatization** Immobilization, reproduction, decontexualization, and distribution of messages	**Digitization** Production, distribution, and interaction in context Control of messages on a bit-by-bit basis
Regulation of human groups	**Organicity** The members of an organic group share the mutual awareness of their identities and acts.	**Transcendence** The members of a molar group are organized into categories, united by leaders and institutions, managed by a bureaucracy or held together by enthusiasm.	**Immanence** A large, self-organizing community is a molecular group. Using every resource of micro-technology, it enhances its human wealth, attribute by attribute.

controls on a much finer level of detail. It will avoid mass production. Ultrarapid, precise, acting on the level of microstructure, molecular technology from cold fusion to superconductivity, from nanotechnology to genetic engineering, will reduce waste and scrap to a minimum. The political engineering suggested here must become part of a vast and profound technological movement toward microstructure, which will include other types of engineering, technologies other than those that are specifically human. This trend will become most obvious in three areas: the control of life, matter, and information. After examining the rough outlines of the evolution toward molecular technologies in these fundamental sectors of human activity, I will show how political technologies, those that organize and legitimate communities, might take advantage of these approaches without losing sight of their own unique ends.

Life

Natural selection could be considered a technology that life applies to itself. Nature, a pure "will to power" that operates beyond the molar or molecular level, shapes and preserves species without any predetermined goal, without reason. On the human level its effects take place with infinite slowness.

Artificial selection represents the next most important biological technology. Using the same basic processes as natural selection, it innovates by *finalizing* and *accelerating* the formation of species. It was at the moment of the great Neolithic revolution (one that lasted several millennia) that mankind began to deliberately select and domesticate plants and animals, and create new species: wheat, barley, rice, corn, dogs, sheep, cattle, poultry, etc. Yet, although the process is much more rapid than natural selection, it still requires several generations for its effects to be felt. On the level of the population as a whole, through cross-breeding and the selection of breeders, artificial selection could only indirectly control the characteristics of living beings, in a quasi-statistical manner. Slow and imprecise, it remains a molar technology.

Molecular biology has opened a path to technologies that can be used to control life on a fundamental level, which governs the form and function of organisms. Even though it is still impossible to directly control characteristics, we are at least beginning to understand how to manipulate genes directly. In principle it will be possible to produce a new species in a few days. The creation of a species or a race depended on geological (natural selection is measured in thousand-year units), then historical (artificial selection takes place in several generations) time periods, and is now beginning to take place in real time, immediately (biotechnology is measured in terms of man-months, hardware, and dollars). Biotechnology is molecular not only because of the scale on which it operates, but also conceptually, because it promotes an operational model of life and its products based on a detailed understanding of the microstructures of organisms and their mechanisms of reproduction, because it can precisely target receptors on molecules, because it can manipulate interfaces and specific features in great detail, as well as reaction dynamics in microscopic networks. The slogan of contemporary biotechnology could be: gene by gene, or molecule by molecule.

Matter

We can classify the technologies for controlling matter into three major categories: mechanics, heat, and cold. Based on the actual data of prehistoric archeology, the mechanical technologies were the first to appear. Evidence for this is supplied by the silex choppers, bifaces, blades, and scrapers that preceded the mastery of fire by several hundred millennia. Mechanical technologies control the point of application of human, animal, or natural forces (tools, weapons, tilling devices, sails, etc.), the transmission of these forces (wheels, pulleys, shafts, gears), and the basic assembly of materials (knots, textiles, primitive architecture, etc.).

The technologies of heat produce energy and transform the fundamental qualities of materials. Closely based on the knowledge of lighting and maintaining a fire, cooking is unquestionably the oldest of the technologies of heat. Beginning in the Neolithic period, pottery and metallurgy appeared. But the great development in thermodynamic technology dates back to the industrial revolution at the end of the eighteenth century, with the increased use of steam engines and the appearance of the civilization of carbon and steel. For the past two hundred years, the chemical industry has been the greatest consumer of the technologies of heat. It obtains its finished products through heating, mixing, and bulk reactions. It transforms streams of raw materials, statistically, externally. The industries based on thermal technologies for processing matter are generally very costly in terms of energy, generate pollution, and are large producers of waste. The technologies of heat are molar.

In the material world the technologies of cold are still in their infancy. We are making progress with emulsions, surface chemistry, crystallography, advanced ceramics, "intelligent" materials in general. We are just beginning to industrialize the most recent discoveries in the physics of solids. In the future, ultrafine technologies of matter will carry out the classic operations of mechanical technology—controlling transmission, point of application, and assembly—with one important difference. Conventional mechanical processes will be applied at the atomic and molecular level, and in such a way that their effects will be similar to those of the earlier technologies of heat. They will modify the innermost properties of materials. But now changes will take place practically without defect, with a minimum expenditure of energy; they will occur at a specific site and within a specified period of time. At the crossroads of physics, chemistry, and materials science, nanotechnology appears to provide sufficient control of the microstructure of material to equal, in granularity of detail, the performance obtained from information technology and biotechnology in their respective fields, and which could even lead to some form of reciprocal interaction between them.

The extreme precision of new forms of chemical engineering, molecular targeting on a mass scale, the possibility of industrially manufacturing macromolecules atom by atom may, within a few decades, eliminate the earlier technologies of heating and mixing used by industrial chemistry. Some contemporary researchers[1] have, in fact, predicted the development of nanosensors, nanocomputers, and nanorobots, operating on a molecular level, that will provide the materials of the future with a form of distributed intelligence, the capability of autonomous production and reproduction, and the programmed response to changes in their environment. Are we able to conceive of how such intelligent materials will be used in the future? The change in material civilization announced by nanotechnology could lead to a reassessment of economic, social, and cultural issues of enormous scope, next to which those associated with the appearance of information technology would appear negligible.

Information

The technologies used to control message flow can be classified into three main groups: somatic, media, and digital. Somatic technologies imply the effective presence, commitment, energy, and sensibility of the body for the production of signs. Typical examples would be the living performance of speech, dance, song, or music in general. While difficult to distinguish from the overall situation in which it occurs, a somatic message is essentially multi-modal. Speech is accompanied by gestures and facial expressions; dance is only truly visible within its container of sound. A somatic message is never exactly reproduced by somatic technology. Whether associated with tradition or lineage, it is always unique because it is inseparable from a changing context. Depending on the situation and the adjustment of his intention, the producer of a somatic message modulates, adapts, and continuously varies the flux of signs of which he is the source.

Media technologies (molar) focus and reproduce messages to ensure they will travel farther, and improve distribution through time and space. To the extent that they produce durable or transportable semaphores, the creation of statuary, jewelry, paintings, or tapestries are already protomediacentric activities. Messages continue to be transmitted in the absence of the living body of their creators. The transition to media properly speaking (that is, mass media) occurs with the introduction of technologies for the reproduction of signs and marks: seals, punches, molds, dies for coining, etc. Writing, in its ideographic manifestation (the notation of an idea by means of a conventional image), is related to drawing and therefore to protomedia. But in its systematic aspect, as something codifiable and reproducible, media has tended, ever since its inception, to become all encompassing. The alphabet, when coupled with reading and writing, becomes an element in a mechanism for reproducing speech. It is the first technology for recording and playing back sound. By enabling the mass reproduction and distribution of text and image, printing inaugurated the age of the media. This development reached its apogee between the middle of the nineteenth and twentieth centuries, through the use of photography, sound recording (voice printing, record players, and tape players), the telephone, film, radio, and television.

The media focus, reproduce, and transport messages on a scale that somatic methods could never achieve. But in so doing, they decontextualize them and cause them to lose their ability to adapt, which they possess when issued by a living being. With the partial exception of the telephone (although only a single recipient is involved), the mediated message ceases to interact continuously with the situation that gives it meaning. Because of this the signification carried by such messages is greatly reduced when they reach the majority of their recipients. From the receiver's point of view, we can overcome this by some form of hermeneutic activity (reading as an intensive effort of recreation). From the sender's point of view, we can limit ourselves to the smallest common denominator of those who constitute the "public." The media are molar technologies; they act on messages externally, in the mass.

Though they have a strongly retroactive effect on message production, conventional media are not, as a first approximation, technologies for sign creation. They are limited to the fixation, reproduction, and transportation of somatically produced messages. Engraving records a gesture; photography and film are based on the optical capture of material situations, painted sets, or the interactions of actors; prior to the recording process, musicians make their instruments vibrate, singers give music a voice. But this simplistic division between original and reproduction has been explicitly contested by the cinema since its origins. For although the raw image or sound may be stored on the recording, the global message—the film—results from a process of montage. The media, a

recording and distribution technology, thus retains a generative potential that challenges any straightforward relationship between an "original" somatic message and
its mediatization. By the end of the sixties, some studio
recordings were so dependent on the technologies of
amplification and mixing that it became impossible to
reproduce them during a live performance. Writing, the
archetypal media, has always depended on techniques of
montage, mixing, and spatial orientation. Writing provides the semiotic foundation for *sui generis* modes of
expression and communication, which are far from limited to the simple reproduction of speech.

In this sense digital technology has always haunted
the media. Digitization is the absolute of montage, montage affecting the tiniest fragments of a message, an indefinite and constantly renewed receptivity to the combination, fusion, and replenishment of signs. Using the
terminology of digitization, we would speak of computation, calculation, or information processing, rather than
montage. Although accelerated by the machine, the ancient operations of writing still remain. Information technology is molecular because it does not simply reproduce
and distribute messages (which it does better than conventional media), but enables us to create and modify
them at will, provide them with a finely graduated capability of reaction through total control of their microstructure. Digitization enables us to create, modify, and
even interact with messages, atom of information by
atom of information, bit by bit. We can retain the timbre
of a voice or an instrument, while using it to play a
different melody. Accelerate the rhythm of a piece without ending in shrillness. Transform the color of a flower in
every single frame of a film. Increase the size of an object
by one hundred twenty-eight percent while retaining its
shape. Recalculate the perspective of a landscape when

our point of view shifts nine degrees to the left or right ... down to the second.

In traditional written communication, all the resources of montage are employed at the moment of composition. Once it is printed the physical text retains a certain stability, aside from any potential disorientation or reorientation experienced by the reader. Digital hypertext automates and materializes the operation of reading, and considerably magnifies its scope. Always in the process of reorganization, it offers a reservoir, a dynamic matrix through which a navigator, reader, or user can create an individual text based on the needs of the moment. Databases, expert systems, spreadsheets, hyperdocuments, interactive simulations, and other virtual worlds are so many potential texts, images, sounds, or even tactile qualities, which individual situations actualize in a thousand different ways. Thus digitization reestablishes sensibility within the context of somatic technologies while preserving the media's power of recording and distribution.

But this occurs only when the potential of information technology is effectively exploited. A compact disc containing recorded music is a form of mass media rather than a molecular technology, even though it is digitally encoded. Multimedia is still just another form of media. Editorialized on CD-ROM, a hyperdocument, even though it bears some of the characteristics of "interactivity" inherent in digital technologies, provides us with less plasticity, dynamism, or sensitivity to the evolution of context than a hyperdocument that has been enriched and restructured in real time by a community of authors and readers on a network. The molecular treatment of information creates a cyberspace that virtually interconnects all digitized messages, multiplies sensors and semaphores, generalizes interaction and calculation in real time. Cyberspace tends to reconstruct, on a larger scale, the unbroken plane, the *continuum indivis*, the living and changing bath that united signs with living beings, as it did signs with signs, before the media isolated and immobilized messages.

Every molecular technology conditions the others. Without the advances made by information technology, genetic engineering would certainly not be as advanced as it is, and it is unlikely that nanotechnology would even have been developed. Yet the technology and science of materials in large part control the progress made by digital technology.

Human Communities

Molecular technologies, operating on the micro level and at ambient temperature, can be contrasted with molar technologies, which are bulk operations requiring heat or cold, and with age-old processes that indistinctly targeted entire populations, were slow to reorganize because of indiscriminate methods of selection and mixing, required heat and massive amounts of radiation, and generated considerable waste and scrap. In learning to control life, we tend toward finely granulated modes of action, those that are targeted, precise, rapid, economic, qualitative, discrete, calculated, and carefully implemented at a specific moment in time, while closely following the continuous evolution of goals and situations. I am proposing a similar evolution in the conduct of human affairs.

Is such a thing possible? The advance of molecular technologies for processing matter promises an unprecedented increase in the productivity of human labor, an acceleration in the rate of economic mutation. Should we continue to subordinate social identity and the psychological survival of the individual to forms of labor (notably salaried employment) that stabilized during the nineteenth century, at the height of molar technology? The possibility of cyberspace allows us to envisage forms of economic and social organization based on collective intelligence and the enhancement of humanity in all its variety. Yet we continue to focus our attention on such things as capturing the multimedia market. We have attained an unprecedented degree of precision and accuracy, are capable of great economy in processing signs and objects, yet show little concern for systematizing and extending equitable methods of interaction and relation where human beings are involved. Do new social conventions exist that would prevent us from wasting our skills, from wasting any human quality in general? How do we stop treating men and women entropically, in bulk, in the mass, as if they were interchangeable within their category, and consider them as unique individuals? How can we make it obvious that the other is a unique bearer of skill and creativity? When intelligently managed organizations are no longer able to confront the complexity of existing situations, how do we make the transition to organizations that are collectively intelligent? These are representative of the problems that face us when we attempt to engineer the social bond—the molecular technology of a politics that has yet to be invented.

We can distinguish three main ideal types among the variety of political technologies. Families, clans, and tribes are *organic* groups. Nations, institutions, religions, large corporations, as well as the revolutionary "masses" are *organized* groups, molar groups, which undergo a process of transcendence or exteriority in forming and maintaining themselves. Finally, *self-organized*, or molecular, groups realize the ideal of direct democracy within very large communities in the process of mutation and deterritorialization.

No organic group can exist unless each of its members knows the names of the others. Within a collectivity of this type, individuals obey rules, follow traditions, and respect codes. And yet, the organizational principles are not static, reified, or external to the group since they are maintained by the community as a whole. When a member of an organic group carries out an action, the others are immediately able to gauge how that action affects their own situation. Individuals are more or less aware of their actions as a group. Each of them can interact with the others, without recourse to experts in mediation or organization. The majority of the lasting examples of direct democracy are based on organic groups.

The group relies on political technologies of transcendence when it becomes too large for individuals to know one another by name or comprehend in real time what they are doing as a group. At this point leaders, chiefs, kings, and various representatives unite, polarizing the community space. Institutions create temporal continuity. Bureaucracy becomes a separate organ for managing and processing information. A strict division of labor, especially the division between design and execution, is supposed to ensure the optimum coordination of activities. The technologies of transcendence pass through a center, a high point, and from this external vantage, separate, organize, and unify the community. Society remains at the stage of molar technology, however, for in order to comply with the need to manage the mass of humanity, individuals are not considered in terms of what they are in themselves or their relationship to the whole (they are nameless), but as members of a category (caste, race, rank, grade, trade, discipline) within which they are interchangeable. Based on these identities of adhesion, individuals are seen as a mass, as numbers, independent of their molecular wealth. The molar group organizes a kind of human thermodynamics, an exteriorized channeling of behavior and character that squanders individual qualities. Industrialized labor, military service, media audiences, unemployment, poverty, marginality, madness, minorities (or even majorities for that matter), oppression—it's only a short step from mass processing to waste, scrap, and rejects. Transcendence and separation are also molar technologies of heat or refrigeration because, within the groups they organize, change is costly, sudden, and often catastrophic, resulting in coups, revolutions, and uprisings. During these periods of violent transition, periods of revolt and enthusiasm that often give rise to the emergence of charismatic leaders, the group begins to fuse. It becomes a source of energy, exploited by specialists in mediation.

In a system organized around molecular politics, groups are no longer considered as sources of energy to be exploited for their labor but as collective intelligences that develop and redevelop their projects and resources, continuously refine their skills, and attempt to enhance their individual qualities indefinitely. Able to reorganize itself in real time, minimizing delays, deadlines, and friction, the molecular group evolves at room temperature, without sudden change. The politics of separation and transcendence is to the diversity and richness of human action what heavy industry was to natural resources and the environment: It exploits without consideration and in the end destroys more than it creates. When observed on the molecular level of individual lives and human relations, order in molar groups appears to be nothing more than disorder and mindless waste. On the other hand, by generalizing the concept of "zero contempt" characteristic of new methods of management, molecular politics, or nanopolitics, enhances the very substance of social relations at the finest level of detail and on a just-in-time basis. It makes use of every human act, enhances individual qualities. It engineers a social bond that integrates and creates synergy between creativity, the capacities for initiative, and the diversity of skills and individual qualities, without circumscribing or limiting them through the use of categories or *a priori* molar structures. The goal of such a micropolitics is not to model the community according to a preestablished plan. This would obviously mean a return to the worst aspects of mass technology. Rather, it brings into being an immanent social bond, one that emerges through one-to-many relations.

The multiplication of molecular communities assumes the relative decline of media-based communication for the benefit of a cyberspace that is receptive to collective intelligence, a space that becomes increasingly navigable and accessible as molecular technologies become operational and available at low cost. Real-time, large-scale collective intelligence thus requires a sufficient technical infrastructure.

Should some source of power gain control of individuals and communities "quality by quality," society would revert to a form of molar technology; it would again undergo some form of transcendence. Every operation carried out on an external object by an impending subject, within the context of the molecular technologies of matter or message, becomes an implied project, a self-directed action, a form of unfettered reciprocal interaction in engineering the social bond. Within the human sphere molecular technologies provide groups and the individual with the instruments for selectively *enhancing themselves*, quality by quality. Such technologies promote mutual recognition and synergy among anthropic qualities. In the language of material technology, we speak of controlling microstructures, but by making use of molecular technology, we translate this idea into the idiom of self-reflection, subjectivity, and human respect. It is an invitation to the active expression of singularity, the systematic regeneration of creativity and skill, the transmutation of diversity into sociability.

Just as a brain thinks without a center or superbrain to direct it, a molecular group has no need for transcendent mediation to join together. Technical evolution has made transcendence obsolete. Just as nanotechnology can build molecules atom by atom, nanopolitics cultivates its communitarian hypercortex with the greatest attention to detail, the greatest precision and individualization, by promoting the complex interaction of cognitive abilities, fragile sources of initiative and imagination, quality by quality, without any loss of human wealth. Just as messages in cyberspace interact with one another across an unbroken deterritorialized plane, the members of a molecular community communicate laterally, reciprocally, outside categories and hierarchies, folding and refolding, weaving and reweaving, complicating the great metamorphic fabric of their peaceful cities.

4

The Dynamics of Intelligent Cities

A Manifesto for Molecular Politics

Will government be possible during a period of accelerated deterritorialization? The creation of new methods of political and social regulation are now among society's most urgent tasks. Such methods are morally desirable to the extent that they deepen our sense of democracy. They also condition our well-being as a society by helping us resolve some of our most serious and complex problems. My approach is a utopian one, based on a form of direct, computer-mediated democracy—a virtual agora—that is more appropriate than current representative systems in helping us cross the turbulent waters of anthropological mutation.

Technology and Politics

Communication infrastructures and intellectual technologies have always been closely tied to economic and political forms of organization. There are a number of well-known examples of this. The birth of writing is associated with the first bureaucratic states, which were based on a top-down hierarchy, and the first forms of centralized economic administration (taxes, the management of large agricultural domains, etc.). The appearance of the alphabet in ancient Greece is contemporary with the emergence of money, the city state, and especially the invention of democracy. The growth of reading enabled more people to understand the laws and discuss them. Printing made it possible to distribute books over wider areas and led to the development of newspapers, the basis of public opinion, without which modern democracies would never have come into being. Printing was also the first industry of mass production, and the technical and scientific development it promoted was one of the engines of the industrial revolution. The audiovisual media of the twentieth century (radio, television, records, films) have contributed to the emergence of a society of the spectacle, which has profoundly altered traditional methods of operation in cities and markets (publicity).

The close interaction between communications technologies and government structures has been confirmed by recent political events. Well adapted to unidirectional, centralized, and territorialized media, authoritarian governments have been unable to prevent the growth of telephone networks, television satellites, facsimile and photocopy machines, and generally any instrument that stimulates decentralized, lateral, and non-hierarchic communication. Contemporary mass media distribute a range of ideas and representations, which challenge rigid methods of organization and closed or traditional cultures. In spite of the inevitable reaction and temporary resurgence of antiquated social structures, they have demonstrated immense critical power. But although they are able to propagate emotion, display images, and dissolve cultural isolation, the mass media are of little use in helping people think as a group and collectively develop solutions to their problems. Cooperative apprenticeship and the reconstruction of the social bond are achievable with modern communication tools. Now that our societies have felt the critical and deterritorializing powers of conventional media, is there anything to prevent them from experimenting with such tools to develop a collective intelligence?

Technical innovations open up new fields of possibility, which society either ignores completely or embraces without the assumption of any mechanical predetermination. A vast political and cultural plain stands before us. We have an opportunity to experience one of those rare moments when a civilization deliberately invents itself. But this opportunity won't last for long. Before blindly stumbling into a future from which we cannot return, it is essential that we begin to imagine, experiment with, and actively promote, within this new communications space, organizational structures and decision-making styles that are oriented toward a deepening of our sense of democracy. Cyberspace could become the most perfectly integrated medium within a community for problem analysis, group discussion, the development of an awareness of complex processes, collective decision-making, and evaluation.

Inadequacy of Government Structures

Current forms of government stabilized at a time when technical, economic, and social changes occurred much more slowly than they do today. The major political problems of the contemporary world involve disarmament, ecological equilibrium, changes in the economy and the nature of work, the development of the southern hemisphere, education, widespread poverty, maintenance of the social bond, etc. No one has any simple, definitive solution that will resolve these problems. A serious approach to such issues will probably involve the mobilization of diverse skills and the continuous processing of enormous streams of information. Moreover, the problems in question are all more or less interrelated within a globalized space. Ultimately, their resolution will require negotiations among a large number of agents, which vary in terms of their size, background, and short-term[1] interests. None of our contemporary systems of government were designed to respond to such needs.

Today's decision-making and evaluation procedures were established in a world that was relatively stable and had a simple communications ecology. Information today, however, is torrential, oceanic. This divergence between the flood-like nature of information flow and traditional methods of decision-making and orientation is becoming greater and greater.[2] By and large, systems of government still use molar technologies of communication. Administration is most frequently based on traditional management techniques, which are slow and rigid, implemented through static forms of writing. Information technology is used only to rationalize and accelerate bureaucratic performance, rarely to experiment with new forms of organization or innovative, decentralized, flexible, and interactive methods of information processing. Politicians have found that the communications and intellectual spaces in which they operate have become increasingly polarized by the mass media, primarily newspapers, radio, and television.

To respond to the accelerated rate of change, the wide-scale use of digital simulation technologies and real-time information access, along with interactive forms of communication, could prove extremely useful if available to every citizen. How will we be able to process enormous masses of data on interrelated problems within a changing environment? Most likely by making use of organizational structures that favor the genuine socialization of problem-solving rather than its resolution by separate entities that are in danger of becoming competitive, swollen, outdated, and isolated from real life. The cooperative parallel processing of problems will require the design of intelligent tools for filtering data, navigating within the information stream, and simulating complex systems; tools for lateral communication; and the mutual recognition of individuals and groups on the basis of their activities and skills. It is likely that some of the emerging technologies for the interactive construction and visualization of the spaces of signification will enable us to move in this direction. The generalized use of such virtual agoras will noticeably improve the formulation of questions and the negotiation and decision-making process within heterogeneous and dispersed communities.

The mobilization of social skills is inherently technological and political. The only way we can improve democracy is by fully exploiting contemporary communications tools. Conversely, a deepening sense of democracy as a form of collective intelligence could become a goal that is both socially useful and gratifying for the builders of cyberspace. The most socially useful benefit of computer-mediated communication will be to provide people with the means to combine their mental forces in constructing intelligent communities and real-time democracies.

Will the Virtual Agora Be Limited to an Elite?

One objection to such a technopolitical suggestion might be that the tools for navigating cyberspace would be too expensive and difficult to manipulate. The electronic agora, it is claimed, would be an elitist luxury, reserved for the wealthiest and best educated. This argument is easily refuted, however. As far as cost is concerned, such a system could be based on the existing material infrastructures, even without the use of costly fiber optic cables. The developments needed to improve data compression and decompression systems and to design communications, navigation, simulation, and visualization software would be minimal compared to the amounts of money spent on defense or the construction of empty office space. No specific investment would be required to develop the terminals. We could make use of currently available multimedia computers. With respect to the barriers to use, contemporary digital tools are becoming less and less difficult to operate. A growing proportion of the population makes use of computers in its work and knows how to interact with one or two software applications. Moreover, the difficulties of learning how to operate a computer would appear to be nonexistent for the younger generation. (What seems like science fiction to those over forty could be quite commonplace within thirty years.) Remember that I am referring only to the use of digital tools for communication, not how to build or program them. As a comparison, it is worth recalling that universal suffrage assumed that citizens were literate. Yet the process of learning to read is very slow, requiring three to four years (sometimes longer) of hard work in specialized institutions, is extremely expensive for the community (schools), and unfortunately still remains inaccessible to some people. Should we thus reject universal suffrage on the pretext that it might be restricted to a prosperous and literate elite? On the contrary, we consider universal suffrage, along with equal access to education, as a right. The basic ability to navigate cyberspace will probably be acquired in far less time than it now takes to learn to read and, like literacy, will be associated with a number of social, economic, and cultural benefits in addition to citizenship.

The telephone and television are currently part of the standard equipment found in households throughout the industrialized world, even among those of modest means. The television serves as the terminal in a communications system designed around a one-to-many star configuration. The message leaves a central point and terminates in a series of scattered receivers. The telephone functions as the terminal in a communications system designed around a one-to-one network configuration. Contacts are interactive but take place only between two users (or a small number of people) at the same time. It is not unreasonable to assume that in a few years' time the majority of households could also be equipped with terminals (cybergates[3]) that are part of a communications system designed around a many-to-many *spatial* configuration. Citizens could thus participate in new methods of sociotechnical interaction, enabling large communities to communicate with themselves in real time. Cooperative cyberspace must be designed as a form of public service. This virtual agora will facilitate navigation and orientation within the knowledge space. It will promote the exchange of knowledge and favor the collective construction of meaning. And it will make possible the multicriterial evaluation in real time of a dizzying number of proposals, information sources, and processes. Cyberspace could become the locus for a new form of direct, broad-based democracy.

Representative and Direct Democracy

We cannot respond to the unprecedented problems now facing us on the basis of historical experience or tradition. Obviously there was no opportunity for political philosophy to analyze and discuss direct real-time democracy in cyberspace since the technical possibility didn't exist until the mid-nineteen eighties. Athenian democracy brought together a few thousand citizens, who met and discussed issues in a public space that they could walk to. When modern democracies were born, millions of citizens were dispersed throughout large territories. It was therefore *practically* impossible to create direct democracy on a broad scale. Representative democracy could be considered a technical solution to these problems of coordination. But when better solutions appear, there is no reason not to explore them. Classical pluralist and parliamentary governments are obviously preferable to dictatorships, and universal suffrage to censal suffrage. We should not, however, fetishize particular sociotechnical methods. The democratic ideal is not the election of representatives but the greatest participation of the people in public life. The traditional vote is only a means. Why shouldn't there be other means, based on the use of contemporary technologies, which would promote a form of citizen participation that is qualitatively superior to that obtained through the use of ballots dropped into boxes?

Today, aside from coordinated group activities, the effective participation of citizens in public life takes the form of the vote. When the voter adds his voice to a platform, a spokesperson, or a party, he adds a small weight to one side of a balance, makes a minuscule difference in the importance of a proposal. The vote enrolls the citizen in a molar process of social regulation in which his acts have only a quantitative effect. The individuals who have cast the same ballot in the voting booth are practically interchangeable, even though they are faced with very different problems, even though their arguments and positions are subject to a multitude of slight variations. Effective political identity is reduced to our allegiance to a few simple, indeed binary categories. Surveys function roughly according to the same principles. The respondent must answer either "yes" or "no" to simplistic questions that have been formulated by other people, and his responses are ultimately used only for statistical purposes. Aside from the fact that the means of expression implemented by the vote is extremely imprecise, it is also discontinuous and allows citizens very little room for initiative. And major elections generally take place no more than every four or five years.

A real-time mechanism for direct democracy in cyberspace would allow everyone to help develop and refine shared problems on a continuous basis, introduce new questions, construct new arguments, and formulate independent positions on a wide range of topics. Together citizens would elaborate a diverse political landscape that was not preconstrained by the gaping molar separation among different parties. The political identity of the citizens would be defined by their contributions to the construction of a political landscape that was perpetually in flux and by their support for various problems (to which they give priority), positions (to which they would adhere), and arguments (which they would in turn make use of). In this way everyone would have a completely unique political identity and role, distinct from any other individual, coupled with the possibility of working with others having similar or complementary positions on a given subject, at a given moment. Obviously some means would be needed to protect the anonymity of political identities. We would no longer participate in political life as a "mass," by adding our weight to that of the party or by conferring increased legitimacy on a spokesperson, but by creating diversity, animating collective thought, and contributing to the elaboration and resolution of shared problems.

The Formation of a Collective Voice

The technopolitical problem of democracy in cyberspace is to provide a community with the means to develop a collective voice without the need for representation. This collective voice could, for example, take the form of a complex image or dynamic space, a changing map of group practices and ideas. Each of us would be able to situate ourselves in a virtual world that the community as a whole helped to enrich and sculpt through their acts of communication. Collectivity is not necessarily synonymous with solidity and uniformity. The development of cyberspace provides us with the opportunity to experiment with collective methods of organization and regulation that dignify multiplicity and variety.

The problem of forming a collective voice is one of the most difficult in political philosophy and practice. Under what conditions can we justifiably say "we"? And what can this "we" legitimately utter as a community, without usurping or reducing diversity? What do we lose in thus referring to ourselves?

When the participants in a demonstration all shout the same slogans, they constitute a communal mechanism of utterance. But for this possibility they pay a considerable price: The shared utterances are few in number and highly simplified; they mask divergence and fail to integrate the differences that individualize us. Moreover, the slogans generally exist before the demonstration takes place. It is rare when individual participants contribute to their creation. The demonstration, like the vote, only enables individuals to construct a political subjectivity by belonging to a category (by using the same slogans or joining the same party). When all the members of a community formulate (or are required to issue) the same proposals, the mechanism of collective utterance operates at the stage of monody or unison. Given the many ways of referring to ourselves, such an impoverished collective voice soon grows monotonous.

Certain forms of organization enable individuals to differentially participate in a final, complex utterance: Books or articles can be written by several authors; film credits can include everyone's contribution to the production, so can plays, newspapers, etc. In the political sphere, the equivalent would be a legislative enactment that has been discussed, modified, amended, and adopted by an assembly. But in this case the utterance results in a finished product that is finalized rather than the open-ended dynamic of voice composition and message negotiation. Such an utterance mechanism is generally dominated by an author, a director, a chief editor, a conductor. But here the means of enunciation is already polyphonic. However, such a symphony is not yet sufficiently vital, plural, or indeterminate. Its harmony has been established at some point of origin in the past, fixed by some point of termination still in the process of development, or managed from above by some "point of transcendence" that directs the entire process. To be completely free, however, the voice of the community should hang on its breath; it should flow ceaselessly and renew itself in real time.

Cyberspace could harbor mechanisms of speech capable of producing living political symphonies, which would enable human communities to continuously invent and express complex utterances, open up the range of singularity and divergence without the need for pre-constrained forms of participation. Real-time democracy tries to construct the richest collective voice possible, the musical model of which would be the improvised polyphonic chorus. For individuals, this is especially difficult because each of them is at the same time called upon to (1) listen to the other members of the chorus, (2) sing slightly out of register, and (3) find a point of harmonious coexistence between his own voice and that of the others, that is, improve the overall effect of the ensemble. Singers must therefore resist three "harmful attractors": the desire to mask the voice of their neighbors by singing too loudly, the urge to remain silent, and the tendency to sing in unison. In this ethics of the symphony, we can recognize the rules of civilized conversation, politeness, worldliness. Ultimately, this means that we refrain from shouting, listen to others, don't repeat what has just been said, answer in turn, and try to say something pertinent or interesting depending on the state of the conversation. Direct democracy in cyberspace would implement a form of computer-mediated civility. This new democracy could take the form of a large-scale collective game in which the most cooperative, the most urbane, the best producers of consonant variety would win (but always temporarily), rather than those most capable of assuming power, silencing the voices of others, or capturing anonymous masses within molar categories.

The capacity for computation, synthetic visualization, and immediate communication characteristic of cyberspace are indispensable to the large-scale operation of such a device for symphonic production or political polyphony. Obviously, the constitution of the social group can never escape the necessity of mediation. My hypothesis is simply that this mediation could be immanent rather than transcendent. In transcendent systems the mediators are gods, myths, hierarchies, representatives. In immanent systems the mediator between the individual and the group is an electronic tool, held by thousands of hands, which continuously produces and reproduces a varied text-image, a cinemap watched by thousands of eyes, structured by ongoing debate and the involvement of all citizens. The role of the virtual agora is not to make decisions for people but to help produce a mechanism for collective utterance animated by living beings. The technical mediator will continuously compute in real time the discourse-landscape of the group in such a way as to minimize its impact on the singularity of individual utterance.

Until quite recently the majority of group mediators were humans, whose role had transformed them into supermen, quasi-deities (kings, heads of state or government, heroes, media superstars), or subhumans, token victims, enemies capable of polarizing the latent violence in a society. Is there an aspect of anthropological destiny associated with heteronomy, transcendence, divinization, or persecution? While they may not eliminate transcendence and heteronomy, new technical possibilities combined with appropriate organizational and legal progress could at least confer upon them the status of unfortunate desuetude, currently ascribed to human sacrifice, slavery, piracy, torture, apartheid, totally planned economies, and dictatorships. What we consider barbaric today was once held to be quite normal, something imposed by human nature, perhaps even desirable. Immanent technical and legal mediation designed to help implement collective utterance may render obsolete certain schools of anthropology that are quick to believe in the eternal necessity of divine or all too human forms of mediation in giving shape to the unity of a group. Who can be certain that victims, gods, transcendent powers, that heteronomy in general is the only voice that will bind a community?[4] And under such circumstances, what would be the self-fulfilling effect of prophecies presented as facts?

Dynamic of the Intelligent City

The intelligent community is the new face of democratic public life. Invested with this ideal, "molecular politics" releases the grip of territorial power, momentarily suspends the activity of the deterritorialized networks of global economy, and initiates, within the vacuum thus created, rhizomatic processes, the folding and refolding of collective intelligence. This does not necessarily imply the formulation of a program, a "content" for real-time democracy, but merely indicates a way of doing things, the description of a few rules in a new game. It is important that collective intelligence not focus on a particular goal or become reified through any of its internal actions, a particular phase of its dynamic, when its essence is autonomous movement, the self-creative process. The goal of the intelligent city (which should be understood as a moral and political entity rather than a physical place) is its own growth, its densification, its extension, its way of folding in on itself, and its openness to the world. From a political perspective the major phases in the dynamic of collective intelligence are listening, expression, decision-making, evaluation, organization, connection, and vision, all of which are interrelated.

Let's enter the circle and begin by *listening*. The intelligent city not only listens to its environment but also to itself and its internal variety. As I have tried to emphasize, post-media modes of communication are capable of restoring the diversity that arises from effective practice. Listening consists in making visible or audible, in coaxing forth the ideas, arguments, facts, evaluations, inventions, and relations used to weave a social reality, a social body, in the very depths of its being: projects, individual abilities, original modes of relation or contractualization, organizational experiments, etc. In a situation in flux, official languages and rigid structures do nothing more than blur or mask reality, lead to our disorientation within it. Increasing the transparency of the social body to itself (rather than the transparency of the individual to power) assumes that we allow the singularities that populate it to express themselves in their own language, to invent their own description and their own projects without imposing some preestablished code on them.

The willingness to listen also implies a return phase, a bouncing back. It assumes a dialog or multilog. Far from being fulfilled through some transcendent entity or limiting itself to the simple, passive recognition of difference, listening is itself an immanent process within the community, a creative circularity. Thus, indicating to the collective that it has been heard by everyone amounts to providing it with the means to understand itself or, rather, to get along with itself. This brings us fairly close to the birth of the social bond: mutual understanding. The means for collective listening are to real-time democracy what the tunneling electron microscope is to nanotechnology. No activity is possible on a micro scale without some form of molecular perception. For this reason I prefer to speak of the molecular receptivity of a community—its ability to listen—as an emerging process, rather than of "communication" or "information," which, to a large extent, connote molar media. The term "listening" is preferable to communication because it evokes the emptiness of a vacuum rather than the fullness of a channel, because it indicates an attention to requests and proposals rather than the supply of information and the juxtaposition of discourse. Listening reverses the direction of the media. It amplifies the many-voiced murmur of the community rather than serving as a conduit for its representative. The media can continue to inform us of disasters and broadcast images of power. Real-time democracy is based on post-media mechanisms, a network of molecular communication, positive practices, resources, projects, skills, and ideas.

Based on such a continuous process of listening, the individuals and groups that guide the intelligent city can express the problems that appear most important for communal life, assume a position concerning those prob-

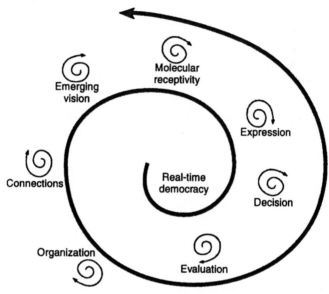

DYNAMIC OF THE INTELLIGENT CITY

lems, and formulate arguments based on those positions. In this way the effective political identity of an individual is no longer characterized by his inclusion in a category but by a singular and provisional distribution across the open space of problems, positions, and arguments, a space that everyone helps shape and reshape in real time. Majorities and minorities must then be viewed as pluralities, for they are no longer associated with a program of molecular government but with emerging and more or less persistent problems. Majorities will only be formed in response to specific questions, on the basis of genuine communal experimentation. Minorities will have the possibility of expressing their views, as long as their proposals do not hinder the operation of real-time democracy or threaten the majority. Assuming they are evaluated, minority initiatives and experiments represent an essential dimension of real-time democracy, for they lead to the exploration of alternative solutions to social problems.

Once decisions have been made and implemented, they will obviously be evaluated in real time by the community based on a number of criteria. The methods of evaluation themselves will, of course, be subject to debate and also evaluated in real time. Real-time democracy maximizes the responsibility of the citizen called upon to make decisions, accept the consequences of those decisions, and make judgments on the basis of those decisions. Evaluations must be seen as taking place within the context of public services or the application of laws. The extension of democracy assumes an enhanced sense of responsibility. It is clear, however, that exposing the collective results of individual and communal decisions is a way of strengthening the attitudes and practices associated with responsibility. Consequently, citizenship cannot be disassociated from an education in citizenship.

The following stage of collective intelligence, organization, consists in distributing functions and entities throughout public life, sharing tasks, regrouping strengths and skills. Organization follows from the actions that precede it. The assignment of responsibilities and resources, if it is to be efficient, that is, capable of stimulating processes rather than simply strengthening territorial space, must of necessity be part of a continuous cycle of listening, expression, decision-making, and evaluation. Isolated from any previous actions, the organization is reduced to artificial separations, lifeless, formal groups, the simple exercise of power. Molecular politics resists the temptation to organize through separate entities. It plunges molar forms of organization into the cycle of collective intelligence. Seen from this perspective, the state and the current structures of government could be maintained, providing their functions are redefined. They would become the guardians, guarantors, administrators, and executants of collective intelligence. The organization would help augment the self-visibility of the social body by making separation and identification easier. In particular, the identification of skill and resource centers plays an important role in improving comprehension and providing direction for citizens. Organization thus favors lateral connections and forms of cooperation, a factor that contributes to its own self-doubt, a state of permanent disorganization. A basic phase in the cycle of collective intelligence, the organization becomes a self-organization, or rather, it appears as the organizing moment of a more universal self-organization.

In the context of the intelligent city, organization cannot be conceived without its disorganizing complement: the lateral connection. The flow of circulation, the folding, refolding, and unfolding of the self in a space where meaning and human relations are in close contact, animates and traverses real-time democracy on a continuous basis. Virtual worlds of shared signification could favor any form of lateral connection and unfettered negotiation without the need for representation. Seen from the outside, this might appear as an instance of disorganization, a purely negative blurring of established differences and boundaries. Such lateral contacts, however, can't be carried out blindly. Endogenous circulation is the immediate consequence of acts of listening, expression, decision-making, and evaluation. It is because social molecules are able to identify one another in their singularity—and because the processes involved have become visible—that unforeseen gatherings, the desire for collaboration, displacement, and exchange, can occur. In this way are the maps of the city rearranged.

The virtual agoras of molecular democracy promote the process of recognition between people and groups, enable them to meet, negotiate, contract with one another. In this sense the development of instruments that will enable us to orient ourselves within the political, social, institutional, and legal complexity will become indispensable, providing they are based on some form of continuous receptivity to the real demands and practices of citizens. One of the goals of real-time democracy is to create a fully transparent market for ideas, arguments, projects, initiatives, expertise, and resources, one in which pertinent connections are established as quickly and cheaply as possible. Untouched by random interactions or Brownian movements, the intelligent city promotes the accurate targeting of molecular reactions and processes. It enhances the human qualities that give it life to as great an extent possible.

The swirl of molecular processes should not block the emergence of a global vision, however. Vision shouldn't be understood here as a fixed image of the future, a potential canvas or mesmerizing sign from another world, but rather as an act of seeing, the flowering of a collective vision, a vision of the self in the process of becoming. This vision is the product of our previous acts: listening, expression, decision-making, evaluation, organization, connection. Permanent retroaction results in the generation of a dynamic. This vision doesn't come from above; it is not the act of an organ distinct from collective intelligence. It emerges from interactions and contacts, circulation, encounters. Vision is the moment when molecular processes delineate or announce a global form, hollowing out a few large attractors. Among the instruments of this vision, the virtual images of this communal dynamic supplied by cyberspace play an important role. These images are synthetic or cartographic, but they can also be endlessly explored, unfolded as a hypertext. They both record the expressions of individuals in the community and enable everyone (individuals, groups, associations, institutions, local communities, corporations) to integrate the shared vision of the community in preparing their future.

Appropriately distributed, global vision is reflected and diffracted in individual projects and strategies, orients and polarizes molecular processes. The unifying vision of diversity only helps animate collective intelligence when it is immediately distributed and autonomously supervised by members of society, thus informing the molecular strategies and practices that in turn help energize the image of community dynamics. Vision is the emerging and global aspect of listening.

Real-time Democracy

The perspective of real-time democracy immediately gives rise to a number of questions, primarily about the unity of the city, the continuity or consistency of its politics. Aren't we risking the loss of memory, erratic changes, uncontrolled crowd movements? Will we still be able to formulate a long-term perspective? Before directly addressing these problems, we need an overview of the important notion of political temporality. This involves two related questions. First, what does it mean to apply real-time methods to social processes in general rather than to information processing? Second, and from a more directly political point of view, what exactly does real-time mean in the expression real-time democracy?

We have seen that molecular technologies are faster than molar technologies. They are carried out in real time, that is to say, they reduce to zero the amount of time before a result is obtained. What are the differences then between real-time processes when applied to computing and transmission and when applied to human affairs? The first difference is quantitative. When communities are involved, real-time processes take place on a very different scale than digital processes. A digital simulation reacts immediately to the change in a variable but an individual doesn't change his mental models and patterns of action quite as rapidly. Groups learn even more slowly than individuals. The second difference is qualitative. For humanity, shortening delays can no longer be seen as an end in itself. The exercise of physical vitality and the enjoyment of human qualities nourishes time, and it would be absurd to want to minimize this. From the point of view of subjectivity, the problem is not to shorten time but to enrich it. If the acceleration of activity leads to the impoverishment of lived time, then in terms of human economy, this represents a loss rather than a gain. By combining qualitative (the subjectivity of time) and quantitative differences, we can understand why the negotiation and acclimatization of novelty within a community occurs slowly. Novelty displaces or disturbs a number of different habits, ways of doing things, identity adjustments, and relational equilibria. Collective apprenticeship is slow because it involves interactions and negotiations among autonomous beings, capable of refusal, each of whom occupies the center of a world. Not as intelligent or free as men, molecules and bits display, in comparison, very little inertia. They can be easily manipulated in real time. The slowness and pace of collective processes are the sign of human nobility. It takes time to learn, to think, to innovate, to make decisions as a group. It requires even more time to form judgments collectively, adjust and deploy languages, knit a community together.

It is important to stress the fact that, within the framework of collective intelligence, real-time democracy is the absolute antithesis of the demagoguery of live action broadcasts and the immediacy of crowd behavior. In fact, two different time frames for collective intelligence need to be distinguished: the time required for their formation and the time associated with their mode of action once formed. The first time frame (or temporizing time frame) is by necessity "slow" and cannot manifest itself in the moment, to the second. When pressed, it balks, simply because it is autonomous. Though it makes use of the molecular technologies and digital networks of real time, it does so to better follow an internal rhythm, one that is subjective, secret, plural, and complicated, that neither clocks nor timetables can measure.

On the other hand, collective intelligence is more "rapid" than organic or molar groups. In fact, what is intelligence, this capacity for learning and invention, other than the ability to accelerate? Inventions almost always help us reach a goal more quickly. *Homo sapiens* created culture, which advances much more rapidly than biological evolution. Technology, language, thought in general, are accelerators. It is for this reason that collective intelligence works as hard as possible to hasten the apprenticeship process, augments its capacity for reorganization, hastens innovation, multiplies the powers of invention. A more intelligent group is also a faster group. But it can only achieve this cognitive speed by mobilizing—and thus respecting—the autonomous subjectivities that it comprises rather than by trying to synchronize them with some external time. The real time of collective intelligence can only be emergent; it synchronizes the intensities of thought, apprenticeship, and life.

We come now to the heart of our problem. There is nothing paradoxical in the idea of a real-time democracy since democracy is by nature something that occurs in real time. As it is most commonly understood, democracy can be contrasted with the arbitrary nature of tyranny or the power of a minority; it posits a system of laws that are valid for everyone and determined by all (or at least by a majority). This implies that the goal of democracy is to realize and preserve the autonomy of a group of citizens: the polis establishes its own laws. But autonomy, as the term is understood today, is incompatible with resignation to a status quo. It assumes an aptitude for change, for questioning, for learning. The autonomous being has the power to escape his past, he refuses to be narrowly determined. A sovereign being, he can modify established laws or introduce others. Yet, following Emmanuel Levinas, transcendence is precisely that which is absolutely in the past, always already behind us, and which we are therefore unable to grasp. When a community decides to establish laws or forms of organization other than those of its ancestors, it escapes the weight of tradition, the hold of transcendence, to focus on the present interests of the community and the new objectives they introduce. The modern polis comes into being as an autonomous entity. Democracy as the quintessential political regime is a "present in the service of a future," rather than a static present dominated by its past or by transcendence (heteronomy). The expression "real-time democracy" is therefore pleonastic, since democracy's essence is the process of collective decision in the present and the permanent reevaluation of the law. To reiterate, if we currently appeal to the deliberation and decision-making of the citizen only from afar, we certainly do not do so on the basis of democratic principles. The periodically renewed delegation of responsibility is a stopgap, representing our inability to bring about an uninterrupted sense of collective intelligence. Because virtual agoras can create spaces for communication, negotiation, the emergence of collective speech, and real-time decision-making, there are fewer and fewer technical arguments for perpetuating the fragmented despotism of delegation.

We can now address any concerns about the absence of a long-term politics and sense of continuity in a real-time democratic environment. It should be pointed out that it is our current governments, that is to say, our elected representatives, that are subject to the short-term and fragmented vision of the media. The absence of long-term vision or politics arises from the close interaction between representation (a form of molar politics) and television (a molar communications technology). The system is such that reelection is the major goal of political representatives, who depend on the instantaneity and absence of memory characteristic of the media. The politics of the spectacle maximally personalizes risk, fascinates citizens, atomizes them, agglutinates them, provides them with little grasp of the operations of political life. It is important that we clearly distinguish between the real-time democracy that could be implemented in cyberspace and the media politics based on the infernal triptych of television, polls, and elections. Real-time democracy does not mean television broadcasts followed by online voting. On the contrary, it is a building block in the laborious but continuous construction of a collective and interactive debate, in which everyone can contribute to formulating questions, establishing positions, proposing and weighing arguments, making and evaluating decisions.

Will such a long-term approach be popular? Will it appeal primarily to those who will determine their collective future and that of their children, or those seeking reelection the following year? Will short-term measures have any appeal? Yes, to those interest groups lacking real decision-making power, who are condemned to making claims but are not subject to any form of evaluation. What about minorities united around a single, compelling issue, which is subject to their evaluation and self-experimentation? A discontinuous politics is born of the infantile relation between irresponsible categories each of which makes claims for itself without reference to the community, and decision-makers who only respond to such claims on the basis of short-term electoral forecasts. In contrast, real-time democracy initiates a period of decision-making and continuous evaluation, during which a responsible community knows that it will eventually be confronted with the results of its current decisions.

Collective intelligence has no relationship to the stupidity of crowd behavior. Panic, collective enthusiasm, etc., are the result of the epidemic propagation of emotion and representation among masses of isolated individuals. The people in a crowd that is in the grip of panic or wild joy are not thinking collectively. True they communicate, but in the minimal sense reflected by the passive and immediate conduction of simple messages, violent sentiments, and reflex actions. The overall effect of individual actions is completely lost on the individuals who constitute the crowd. It is claimed that the only way to make a community less erratic than an atomized crowd is through transcendence (hierarchy, authority, representation, tradition, etc.). But this is false. Technical and organizational mechanisms can expose the dynamics of the collective to everyone involved, enabling individuals to consciously situate themselves within it, modify and evaluate it. Intelligent communities are the direct antithesis of the incoherence and brutal immediacy of crowd behavior, yet do not channel the community into a rigid structure.

Currently, two molar and uniformizing time frames are juxtaposed within the political sphere. The one associated with the politics of the spectacle is discontinuous, fragmented, without memory, unplanned, incoherent. The other, aligned with the temporality of governments and bureaucracies, is slow, conservative, narrowly focused on the immobile continuity and management of its territory, and governed by simply recycling the past. Noise and monotony. Between these two stumbling blocks, real-time democracy must follow and respect the multiple threads of molecular temporality: individuals, heterogeneous communities that intersect one another, problems that seem to follow their own rhythms. Faced with such general instability, real-time democracy attempts to introduce resonance among these rhythms, temporarily harmonize accents and tempi. It expresses a plural and subjective time. We return to the theme of symphonic improvisation: Our voices are now in phase, they respond to one another, give rise to an improbable symphony. Like music, molecular politics is an art of time.

Totalitarianism and the Economy of Human Qualities

Yet once again, and in spite of the preceding arguments, suspicions arise: Doesn't this real-time democracy mask a new form of totalitarianism? If we agree upon the meaning of these words, the above statement is obviously false. Orwell brilliantly elaborated the formula for totalitarianism: "Big Brother is watching you." Media-centric politics simply reverses the totalitarian formula: Rather than organize the constant surveillance of individuals through a national party run by a dictator, it focuses our attention on political celebrities. Everyone watches the same stars: the president, ministers, journalists, "media people." They are the only ones we see and hear. But real-time democracy is organized not around the vision of power over a society (totalitarianism), not around the spectacle of power (the media), but the communication of the community with itself, knowledge of the community's self. Under these circumstances the justification for power is eliminated. For it is precisely when the community is no longer able to recognize itself, no longer controls its own dynamic, is no longer able to produce complex utterances, that power becomes "necessary." To maintain itself, this power must continuously thwart the emergence of a collective intelligence that would enable the community to forsake such power.

Doesn't the idea of engineering the social bond and optimally enhancing human qualities introduce some form of "instrumental reason" (Habermas) into a political sphere in which such calculation and rationality have no place? Doesn't collective intelligence and its virtual agoras represent the subtle but irreversible triumph of *Technik* (Heidegger)? Doesn't any notion involving the political or moral progress of humanity make reference to an Enlightenment philosophy that has long since been refuted, an outdated unifying modernism, and won't it always end up serving one form of imperialism or another (postmodern, *pensiere debole*, common sense, etc.)? Pushed out the door, the suspicion of totalitarianism returns through the window. Today it is extremely difficult to offer a political proposal that isn't cynically realist, or skeptical, or millenarianist.

Real-time democracy is both a special case and the crowning glory of the economy of human qualities. As a result it effectively participates in the enhancement and optimization of individual qualities. Because it can account for the subjective detail of every monad, every individual soul, an intelligent community, like the God of Leibnitz, calculates the best of all possible worlds. The author of the *Theodicy* claimed that the Great Calculator respected individual free will since he intervened only at the beginning, by the overall choice of a world, without interfering in the chains of cause and effect. The economy of human qualities, however, contains no transcendent moment, even if it were to manifest infinite respect for individual liberties. It is a monadology without God. No one holds power. No one possesses absolute knowledge of the whole. Calculation of the best is thus marred by an ineluctable uncertainty—which is all the better. Since we cannot have perfect knowledge of the totality and since it is impossible to predict the future, calculation cannot supply an ultimate plan for what is best, but continuously tracks itself in an indefinite series of approximations, following in real time the arrival of new information and changing situations.

Because of the diversity of human worlds, calculation of the best cannot be aligned with a monodimensional good, one that is molar, solid, and transcendent. A single good for everyone and for all moments (even those of a commercial nature), blocking the emergence of new forms of strength, would obviously no longer be the *good*. Calculation will therefore follow an undefined number of different criteria, and since there are many worlds, there will be many calculations. Thus goals, technologies, skills, plans, tastes, ideas, the unity of meaning, the act associated with a given value in a given community, in a given context, at a given site, and at a given moment, will assume different values at different times and places. It would be preferable to think of a plurality of calculations of the best, permanently changing within overlapping worlds, rather than the final calculation of a universe. This is the major difference between Leibnitz's monadology and the economy of human qualities: The latter does not incorporate any external calculator or powerful computer that determines what is best for everyone. Far from being centralized, its calculations are fully distributed. In fact, there are at least as many simple calculators as there are monads: The calculators are the people themselves.

It is clear that the desire to impose "the best of all possible worlds" can become the pretext for the worst forms of dictatorship. But in this case the horror does not arise from the search for the best, for a concern with optimization, but from the forced, final, and externalized nature of a molar solution, an aggregate solution assumed to be valid for everyone, and thus fatally inadequate for anyone. By restricting freedom, totalitarianism destroys the vitality of being. Moreover, the imposition of a perfect world only characterizes a theoretical totalitarianism or, possibly, a technocracy. True, historic totalitarianism, such as that embodied in fascism, nazism, Stalinism, or Maoism, is not so much characterized by its search for the best for everyone as the invasion of social life by the problematic of power, by the unlimited interjection of the practices of domination and servitude, and by the unchecked proliferation, into the most obscure corners of social life, of chains of dependence, obedience, and submission. So-called totalitarian societies are characterized by the fact that politics, art, science, language, production and exchange, nearly everything that can create a bond, is structured, polarized from top to bottom by hierarchies that are reproduced everywhere with fractal obstinacy, throughout the extent of indefinitely branching networks, for the pursuit and conservation of power. It is for this reason that such societies ended up sterilizing economic, artistic, and intellectual life, and exerted such frenzied energy in wide-scale massacres and genocide. It is for this reason as well that they destroyed any opportunity for an economy of human qualities, which is to say they destroyed themselves. When the criminal activities of the group in power have destroyed civility, the withdrawal of the dominant party leaves in its wake nothing but unchecked banditry and disorder. Ultimately the path to democracy involves a long collective apprenticeship in law, autonomy, reciprocity, and responsibility.

The much criticized Enlightenment, the idea of the moral progress of humanity, never played a significant role in establishing so-called totalitarian regimes. Unscrupulous political gangs arrived to lead the masses and justify the repression (occasionally even in their own eyes), the demands, their destructive folly, by means of nationalist, racist, imperialist, religious, socialist, Marxist, or other theories. No doubt the theories, religions, the powerful images capable of forging identities, are important. But in our attempt to recognize in whose name the crimes of totalitarianism were committed, we seem to have forgotten the nature of the crimes involved and how they were committed. The least we can say is that the effective practices employed by such governments did not exactly satisfy the ideals of humanity's moral progress. For they were unilateral practices of domination, imposition, and levying. They stifled creativity, obliterated differences, and depended on the use of brute force. They were based on contempt, humiliation, the labeling of individuals as subhuman, and on the general devaluation, waste, and destruction of the vitality of being and human qualities. And yes! We are in favor of progress. We harbor ideas about dangerous utopias of reciprocity, exchange, attentiveness, respect, recognition, mutual apprenticeship, negotiation among autonomous subjects, and the enhancement of human qualities. Moreover, such progress, which is not guaranteed by any historical law, depends on the technical, linguistic, conceptual, legal, political, and other forms of cultural equipment we possess. Individual goodwill is not enough. Universal suffrage, for example, is preferable to censal suffrage; commercial freedom is preferable to a toll on every bridge; printed books, personal computers, and telephones open up new opportunities for communication and learning that would be impossible without them. To continue along these lines, cyberspace is opening up immense perspectives for the expansion of democratic practices. Will we be capable of making use of these new possibilities?

We must first stop identifying any idea of humanity's social, moral, or intellectual progress with some dangerous utopia that immediately results in totalitarianism. Either the denunciation of utopias is simply a mask for conservatism, or the logical criticism of utopia culminates in the dismantlement of the destructive mechanisms of transcendence and power.

Power and Strength

Along with the suspicion of totalitarianism, a related criticism sees in the dissolution of power a significant risk of weakening the human groups that engage in real-time democracy. We live in a period of instability and heightened international competition, from both an economic and military perspective. Under these conditions, the self-transparency of the social sphere, the freedom given to minorities to take the initiative and experiment with new methods of regulation, the molecular distribution of decision-making and evaluation, might appear as weak links in the chain.

But in reality, today's winners are those who succeed in mobilizing and coordinating knowledge, intelligence, imagination, and will. The more information circulates, the more rapidly decisions are evaluated. As the capacity for initiative, innovation, and accelerated reorganization can be developed, competition[5] among businesses, armies, regions, countries, and geopolitical zones is increased. Yet power in general has no affinity for real-time operations, permanent reorganization, or transparent evaluation. In general it strives to maintain its advantages and preserve its acquisitions, maintain situations, and block circuits, all extremely dangerous attitudes in a period of rapid and large-scale deterritorialization. Because it provides an education in collective intelligence, because it is capable of mobilizing, enhancing, and making better use of human qualities, real-time democracy is the most appropriate political form for providing the efficiency and characteristic strength needed in the twenty-first century.

Strength makes things possible; power serves to hinder. Strength can liberate; power subordinates. Strength accumulates energy; power squanders it. The technologies of information and coordination have become sufficiently advanced so that the advantages conferred on a given community by a strong authority structure no longer compensate for the waste of human resources and the restrictions on collective intelligence inherent in the exercise of power. To grow strong, a society must begin to divest itself of hierarchies, both internal and external.

Etymologically, democracy refers to the "power of the people." Such a political system is less harmful, not because it gives power to a majority viewed as a unified mass, but to the extent that it mobilizes collective thought in governing public life. It is favored not because it establishes the domination of a majority over a minority, but because it limits the power of government and provides remedies against the arbitrary use of power. Does this mean that such a constitutional form is preferable because it gives power to the representatives of the people? No, but simply to the extent that it replaces specific regulations, privileges, and monopolies with general mechanisms of regulation. We are democrats because democracy limits power to the minimum needed to insure respect for the law.

From the Greeks we inherited a political typology that enables us to answer the question: Who holds power within a political system? But our concern is no longer with giving power to the people, or their representatives, or anyone else. Today, the most pressing political problem is not assuming power but increasing the strength of the people, or groups of people. Power results in loss. A shift has occurred, therefore, from democracy (from the Greek *démos*, people, and *cratein*, to command) to a state of *demodynamics* (Greek *dunamis*, force, strength). Demodynamics is based on molecular politics. It comes into being from the cycle of listening, expression, evaluation, organization, lateral connection, and emerging vision. It encourages real-time regulation, continuous cooperative apprenticeship, optimal enhancement of human qualities, and the exaltation of singularity. Demodynamics does not imply a sovereign people, one that is reified, fetishized, attached to a territory, identified by soil or blood, but a strong people, one perpetually engaged in the process of self-knowing and self-creation, a people in labor, a people yet to come.

5

Choreography of Angelic Bodies

The Atheology of Collective Intelligence

As noted earlier, the idea of collective intelligence implies a technology, an economy, a politics, and an ethics. Before discussing its aesthetics, however, I would like to examine the vacuum left by an atheology to whose silent call the art and architecture of cyberspace respond. Until now the virtual worlds that have been constructed have been simple simulations of real or possible physical universes. I am now suggesting that we design virtual worlds of signification or shared sensation, involving the discovery of spaces in which collective intelligence and imagination will be able to expand. To put this in its proper perspective, I will begin this section by introducing some medieval theological notions of collective intelligence and imagination. I will then show how these notions, once reoriented from transcendence to immanence, can be used to describe the concept of the inverse cathedral, sculpted directly from human intelligence. In this context theology is redefined as anthropology. Our concern remains the fusion of the human with the divine (and what other objective can we assign to an art worthy of the name?) but this time we will enable real and tangible human communities to construct their heaven—their heavens—together, one whose light originates in the thoughts and actions generated here on earth. Theology becomes technology.

The Farabian Tradition

The idea of collective intelligence was first explicitly and rigorously described sometime between the tenth and twelfth century by a line of Persian and Jewish theosophists working within the Islamic community. This work was largely based on a neo-Platonic interpretation of Aristotle. Al-Fārābī (872–950), Ibn Sina (the Avicenna who appears in Latin translations, 980–1037), Abū'l-Barakāt al-Baghdādī (?–1164), and Maimonides (1135–1204) were among the major figures in this tradition.[1]

This tradition is important for several reasons. Central to the anthropology of both Al-Fārābī and Ibn Sina was the idea of a unique and separate collective intelligence, common to the entire human race, which could be thought of as a prototype for a shared or collective intellect. This "collective consciousness" was referred to as the *agent intellect* by these Aristotelian mystics because it was an ever active intelligence, one that constantly contemplated true ideas and enabled human intelligences to become active (and therefore effective) by directing toward them all the ideas that it perceived or contemplated. This shared intellect links men to God, a God who is fundamentally conceived as self-reflexive thought, a knowing divinity and form of knowledge rather than an all-powerful deity, a pure intelligence for whom creativity is only an afterthought. Following Aristotle, Farabian theology was less concerned with the power or strength of God than with His enigmatic manner of thinking, His eternal self-contemplation. By analogy, Farabian theology may provide insight into the idea of collective intelligence and the way in which it contemplates itself by contemplating its world. Al-Fārābī's and Avicenna's theory of knowledge, moreover, was inseparable from their cosmology: The world came into being through a process of perception or contemplation and, consequently, all the celestial hierarchies are implicated in the least act of knowledge. This idea is aptly reflected today in the reciprocal involvement of the world and thought (the cosmos thinks in us and our world is saturated with collective thought), which is a fundamental theme of our meditation on collective intelligence.[2] It should be pointed out that from Al-Fārābī to Maimonides, imagination, far from being discredited, as it was within a Platonic tradition exclusively attached to pure intelligence, was destined to play an important role in thought. For Al-Fārābī and Maimonides, it was the prophets who achieved the highest degree of understanding. Not only did a shared intellect nourish their powers of reasoning (a privilege also granted to philosophers, scientists, lawyers, and politicians), it also made full use of their exceptional abilities to perceive mental images (a power reserved to prophets alone). As in Spinoza's third kind of knowledge, the prophets inspired by this shared intellect "saw" or "heard" truth directly. Perception for them always implied a concurrent act of reason, a fully spiritual perception that was in no way dependent on physical or material processes.

From our perspective at the end of the twentieth century, we can reappropriate this philosophy, which, inspired by Aristotelian and neo-Platonic thought, inherits the Greek hatred of the infinite. God, the angels, thought, and the world are apprehended in *qualitative* terms. God is not infinitely *more* than we are (more powerful, wiser, more just, etc.), but radically different: an absolute unit of thought engaged in self-reflective thinking. Yet because this divinity as "other" is qualitatively finite, we can contemplate its reintegration in the finiteness of a humanity that is ceaselessly becoming other.

The Agent Intellect

In the theology of Al-Fārābī and Avicenna, God does not create the world through some special act of will, there is no *"coup d'état* within eternity" but a series of necessary and eternal consequences of the act of self-reflective divine thought. The world emanates from God as an afterthought because of the overabundance of His intelligence, according to an immaterial causality that the neo-Platonic philosophers referred to as procession or emanation.

Through God's contemplation of His own thought emanates the First Separate Intelligence, or First Cherub. It is referred to as a *separate* Intelligence to emphasize the fact that it is "pure" and not attached to any physical entity. This First Separate Intelligence is engaged in three distinct forms of contemplation from which three consequences follow.

First, it contemplates the principle that is the cause of its necessary existence, which is God. Through this First Intelligence's contemplation of God's thought, a Second Separate Intelligence emanates.

Second, the First Intelligence contemplates itself as a necessary emanation of God. From this contemplation arises the soul that drives the First Heaven.

Third, the First Intelligence contemplates the possibility of its existence in itself, independent of the principle from which it emanates. From this third contemplation, which is the darkest, the lowest, the body of the First Heaven emanates.

In its turn the Second Intelligence, or Second Cherub, (1) contemplates its principle, which is the First Intelligence; (2) contemplates itself as emanation from the First Intelligence; and (3) contemplates itself independent of its principle. From these contemplations arise (1) the Third Intelligence; (2) the soul that drives the Second Heaven; and (3) the ethereal body of the Second Heaven. This process continues through the Tenth Separate Intelligence (in Al-Fārābī and Ibn Sina[3]).

The motive souls, or celestial angels, are characterized by imagination, a pure imagination independent of the senses, which enables them to represent themselves and desire the Intelligence from which they proceed. The love of the celestial angels sets the heavens in motion (from which the expression motive soul), a motion that is eternal since these souls never achieve the Intelligences they desire.

The divine influx, from which the cherubim, angels, and heavens arise, ends by exhausting itself. The process of emanation culminates with the Tenth Separate Intelligence, or agent intellect. This agent intellect is also referred to as "the Angel." When used without any other modifier, the definite article indicates the angel of knowledge and revelation, the angel who intercedes directly with humanity.

Through the Angel's contemplation of itself, independent of its principle, emanates not the rarefied body of a heaven but the distribution, explosion, and opacity of sublunar matter, the coarse substance of our base world.

Through the Angel's contemplation of itself as a product of the Ninth Intelligence emanates not the motive soul of a sphere, a celestial angel, but the multitude of human souls whose thick sensual imagination moves the material bodies.

Finally, the most eminent form of thought accessible to the Tenth Separate Intelligence is obviously the contemplation of its principle (the Ninth Intelligence). From this contemplation emanate all the forms of terrestrial bodies as well as the ideas or forms of knowledge present in those human souls disposed to receive them. The agent intellect is the irradiating source of all the forms and ideas of the sublunar world we inhabit.

Humans are always potential intelligences but they can only act (that is, following Aristotelian terminology, become truly intelligent and knowing) when they are illuminated by the Angel. Intelligible forms stream from the agent intellect, and when they reach suitably receptive souls, they enable the transition from potential (possible) to active knowledge (real).[4] We are thus actively intelligent only because of the agent intellect, shared by all humanity, which is a kind of "collective consciousness." For mankind the greatest happiness is obviously its unity with the agent intellect, the ability to fully and completely capture the angelic emission.[5]

But the process of emanation doesn't end here. The divine influx is received by our rational faculty with varying degrees of intensity. Some individuals receive a superabundance of ideas from the agent intellect. Ideas percolate through their rational faculty toward their spiritual imagination, and they redistribute what they have received by prophesying to other men and women. It is from this prophetic source that knowledge continues to expand "horizontally," from human soul to human soul, until the initial influx is exhausted. Those who lack the gift of prophecy but yet receive the illumination of forms with sufficient strength become teachers, writers, and legislators, by power of reason alone, transmitting in their turn, from one to another, this knowledge of divine origin. Like the prophet, they serve as retransmitters. Others do not receive ideas from the agent intellect with sufficient force to distribute their knowledge, but have enough for their individual perfection. Yet others, like a television whose antenna is poorly oriented, have arranged their soul in such a manner that the Angel only illuminates them at rare intervals, if at all, and although all humans are potentially intelligent, some are incapable of making this intelligence active.[6]

From Angelic to Virtual Worlds

How can these medieval speculations on philosophy and theology help us understand the collective intellect of the future? The agent intellect establishes itself as a transcendent collective intelligence. How can we articulate an immanent collective intellect? As an experiment I will retain the Farabian schema, but reverse its principal terms. The eternal divinity of theological speculation is thus metamorphosed into a desirable possibility lying on the horizon of humanity's future. Within this transformed vision, the angelic or celestial world becomes the region of virtual worlds through which human beings form collective intelligences. The agent intellect becomes the expression, the space of communication, navigation, and negotiation among the members of a collective intellect. From this point on we will have no further need of theological discourse but some mechanism, some way of closely combining technological, semiotic, and socio-organizational elements. Of course this is only an ideal, a goal. Effective realization of this goal will inevitably be imperfect and reversible. However, this shouldn't prevent us from describing the "perfect" mechanism, designed in such a way that it enables collective intelligences to flower with the greatest ease.

Redefined from a human perspective, the angelic regions open the communications space of communities to themselves without the intervention of the divinity, or any form of transcendent representation for that matter (revealed law, authority, or other *a priori* forms imposed from above). Virtual worlds will be instruments of self-knowledge and self-definition for humanity, which can then form itself into autonomous and *autopoietic*[7] collective intellects. In place of the ubiquitous agoras and cosmic simulations, these immanent heavens provide *cinemaps*, dynamic descriptions of the world below, moving images of the events and situations into which human communities are plunged. They are home to the "angelic bodies" (or virtual images) of the members of collective intellects—whether individuals or groups—impelling them toward self-awareness and mutual contact. Synthesizing the complexity and transformations of the terrestrial world, virtual worlds enable intelligences to communicate with one another and assist individuals and groups in navigating collective knowledge. Through these virtual worlds, the world below will continue to proliferate, mutate, and open new pathways of individualization that will, in turn, nurture the "angelic space."

In Farabian theosophy the highest degree of reality lies with God, in His absolute unity, in His pure self-contemplation. While theological discourse conceived of unity as a source, anthropology has grown strong with the many. From the technical and social perspective of collective intellect, reality and richness appear to develop from terrestrial and human multiplicity. Angelic worlds reflect upon collective intellects a brightness that grows increasingly dense, powerful, and illuminating as human knowledge varies, diversifies, and pluralizes itself.

I refer to "angelic worlds" in the plural, since we can obviously conceive of a multitude of agent intellects, just as human communities are organized into collective intellects. It is worth noting that while Al-Fārābī, Ibn Sina, and

Maimonides conceived of a single agent intelligence for all of humanity, Abū'l-Barakāt al-Baghdādī, a Jew who converted to Islam, described, in the twelfth century, a pluralistic version of the relationship between mankind and the angelic world. According to al-Baghdādī, souls are grouped in spiritual families, thus forming different species with a common genus. Our souls originate in different angels, and other angels (as many as there are families of souls) are required to perfect our intelligences. There could even be a different agent intellect (or several) for each human being.

Henri Corbin remarked that a degree of individualism was already implied in Ibn Sina's angelology. For even though there is a single agent intellect, the individual, through its intermediation, is in contact with ideas that emanate from God, independent of any tradition, any church, or any established community. It was this risk of individualism that pushed Aquinas to criticize the Avicennan conception of the agent intellect. Contrary to the author of the *Summa Theologica*, I see no inconvenience in allowing the individual to participate directly in the expression of one or more collective intelligences without passing through "hierarchical channels." Our humanist angelology not only encourages direct contact with collective thought, it also encourages intellectual nomadism. The greater the number of collective intellects with which an individual is involved, the more opportunities he has to diversify his knowledge and desire. Moreover, he is enabled to enrich with his living variety the thinking communities he helps construct. In each virtual world we traverse, we will clothe a different angelic body.

Rather than directing upon mankind the intellectual light that descends from God *via* the heavens and superior angels, the virtual world that serves as agent intellect reflects the brightness that emanates from human communities. Angelic regions of a new kind, virtual worlds will thus emanate collective intellects and depend for their existence on the human communities from which they arise.

The importance and reality of knowledge can no longer be measured solely in terms of its origin, but its degree of acuity, incarnation, and implementation by living beings here on earth. The light of virtual worlds will continue to illuminate and enrich human intelligence, not by enabling it to proceed from potentiality to actuality, since it is always effective, but rather by opening up to it possibilities to which it would not otherwise have access, by making it aware of the knowledge of other intelligences, by providing it with new strength of understanding and new powers of imagination.

Everything that flows from top to bottom in theological discourse should be viewed, within the technosocial system, as flowing from bottom to top. From concrete intelligences and the practices of a multitude of individuals and small groups, a virtual world will emerge that expresses a collective intelligence or imagination. In return, the virtual world will illuminate the individuals and groups who have contributed to its emergence, it will enrich them with its diversity, provide them with new possibilities.

In its humanist version the agent intellect illuminates human souls only because it concentrates and reflects the entire spectrum of light that rises from below.

Enigma and Desire

In the context of theological discourse, illumination comes from above, and the Angel's, or God's, representative distributes it among mankind. According to our humanist vision, the individual who is most capable of accepting the lower world that surrounds him and receiving the teachings (most often silent and unconscious) of other men and things will emit toward the virtual world the richness and diversity he has finally mastered. Theology outlined a unidirectional pattern of distribution, one that descends from above before spreading outward with centrifugal force. Anthropology embodies a form of centripetal circulation that rises from below and showers us with a rain that, even as it falls, anticipates future ascensions of newly concentrated knowledge. Sources spring up here below, are distributed throughout the world and among mankind. Before inseminating the immanent heaven of shared thought, participation in collective intelligence begins by accepting human alterity and embracing diversity. As it develops, thought takes the form of apprenticeship, discovery, encounter. Yet every apprenticeship enriches the collective intellect. The figure of the prophet makes way for that of the explorer, engaged in a constant process of learning and discovery.

For theology, he who refuses to turn toward the forms irradiated by the Angel of knowledge is only potentially intelligent. The soul of the ignorant remains dark and opaque; no idea illuminates it. From our humanist perspective, however, no one is ignorant since every life implies and necessarily constructs the knowledge of a world. The judgment of ignorance arises because we define knowledge to be transcendent. Knowledge comes from God, from revelation, the church, the party, the sect, the university, the school, science, method, experts, elders, leaders, the scriptures, television, from any infallible body or system whatsoever. Knowledge is said to exist as something inherent, autonomous, and not as permanent creation, a process of exploration and collective becoming, the shared ideas of millions of sentient beings, the expression of the diversity of lives and worlds, universally distributed throughout mankind. Any transcendent definition of knowledge must exclude those who refuse to submit to it, or whose form of intelligence isn't recognized. On the other hand, an immanent approach to knowledge—one that is recognized as being present wherever there is human life—excludes no one. Thus, no one is ignorant, that is to say, everyone is actively intelligent. But then, since everyone is always already intelligent, of what benefit is the construction of virtual worlds to express collective knowledge? The light that falls from virtual worlds does not imply that individual intelligences proceed from potentiality to act, but from a specific act to new potentials.

By plunging my diaphanous bodies into the virtual world, I perceive not only that which I already know but also the extent of possible knowledge, which remains foreign to me and may always remain so: the knowledge, ideas, and works of others. My angelic bodies in my virtual world express my contribution to collective intelligence or my individual posture with respect to shared knowledge. Yet this angelic body never attains the full extension of the virtual world that contains it and which is the Angel of the collective self.

Within the space that emanates from the collective intelligence, I thus encounter the human other, no longer as flesh and blood, as social rank, an owner of things, but as an angel, an active intelligence—active for himself but potential for me. Should he ever agree to expose his face of light, when I discover the angelic body of the other, I will contemplate his life in knowledge or his knowledge of life, the projection of his subjective world upon the immanent heaven of the collective intellect. Because I do not know what he knows, because our becomings differ, he is, within this space, a unique, incomparable figure of desire: His angelic body reveals this to me as enigma and alterity. It is in this way that the other world, or mystery, of theology becomes the world of the other, or enigma, of anthropology.

Among the medieval philosophers love rose from the soul toward the superior intelligences. In our humanist system it is by their passage through virtual worlds, by acquiring an angelic body, that souls can best imagine humanity, from which perhaps follows, along with the desire of learning, the growth of friendship among men. When referring to those who cease learning, we no longer speak of ignorance but of closure, a delayed life, a rigidity that is impermeable to the proliferation of strength, a refusal to encounter the other as angel, a fear of enigma and desire.

The Problem of Evil

We have seen that in the theological philosophy of Al-Fārābī and Ibn Sina, obscurity and matter, or evil, occur because intelligences contemplate themselves as existing independently of the superior principle from which they emanate. Men forget to turn toward the agent intellect, the separate intelligence neglects the superior intelligences, the First Separate Intelligence contemplates itself without God.

What is the cause of evil from the present humanist perspective? Collective intellects may be tempted to consider virtual worlds as realities in themselves, forgetting the living human beings from which they arise and of which they are merely the expression. This is their dark side. Any illusory autonomization of the figure of the community, any idolatrous fixation on its countenance, any transcendent becoming of the knowledge space will give rise to evil. In such a case, the question of truth is substituted for the uninterrupted dynamic of learning and exploration. Mystery and terror replace enigma and

desire. Exclusion succeeds mutual recognition. And ultimately, forgetfulness of the living and present origin of
virtual worlds, their reification, their separation from the
innumerable human sources from which they originate
will inevitably introduce the obsessive problem of power
within a space where it does not belong. We can then
pose the following absurd question: Who will control
such virtual worlds? This is like asking who will speak in
the name of the community within these virtual worlds,
when such worlds are precisely the very means of self-
organization, self-definition, and autonomous self-con-
struction of the community within the knowledge space.

Any assumption of control by a small splinter group,
any petrifaction of living, collective expression, any evolution toward transcendence immediately annihilates the
angelic nature of a virtual world, which falls at once into
the obscure regions of domination, power, belonging,
and exclusion. I am aware that many claim that self-
organization is impossible; they cannot conceive of a
space in which the question of power doesn't arise, they
consider knowledge a territory to be divided up or a
network to be controlled. I can only encourage them to
continue their explorations in an attempt to enlarge their
subjective world.[8] The angels of the living unite to perpetually form and re-form the Angel of the collective,
the moving and radiant body of human knowledge. The
Angel does not speak. It is itself the aggregate voice or
choral chant that rises from an acting and thinking humanity.

The Intellect, the Intelligible, the Intelligent

It is possible that the agent intellect will encompass
only the space of language in general, the world of signs
that emerge from the human adventure, which we help

enrich and transmit through our life and our acts. Even if we limit ourselves to this weakened and rather banal interpretation, it remains true that we think by and in angelic illumination.

Yet, I wish to defend a strong version of the theory. My hypothesis is that it is both possible and desirable to construct technical, social, and semiotic means that will effectively incarnate and materialize collective intelligence. It would perhaps be simpler to limit theory to suggestive analogies and illuminating metaphors. But when I speak of virtual worlds, I am specifically referring to vast digital networks, computer memories, interactive multimodal interfaces, quick and nomadic, which individuals can easily appropriate. This implies a relationship to knowledge that will differ from that which exists today, the inauguration of an unmediated communications space, a profound renewal of human relations, both within the context of work and within political life, a reinvention of democracy—all possibilities that are embodied in the ideal of the collective intellect.

The collective intellect is a kind of corporation in which each shareholder supplies as capital his knowledge, experience, and his ability to learn and teach. The intelligent collective neither submits to nor limits individual intelligences, but on the contrary exalts them, fructifies and reinvigorates them. This transpersonal subject is not merely the sum of individual intelligences. Rather, it gives rise to a qualitatively different form of intelligence, which is added to personal intelligences, forming a kind of collective brain, or hypercortex.

Yet wasn't this different and entirely other intelligence, which continues to illuminate and exalt, initially conceived as divine intelligence? Doesn't the construction of collective intelligence represent, for human communities, a secular, philanthropic, and reasonable means of achieving divinity? On the condition, of course, that God is stripped of His necessary and transcendent nature and redefined in immanence and virtuality.

I would, therefore, like to continue this preliminary exploration with a discussion of some theological concepts based on Aristotelian and neo-Platonic thought.[9] As we have seen, according to medieval philosophy, man's intelligence only becomes active intermittently. God, however, is eternally active thought because He always realizes, outside time, the perfect unity of the intellect, the intelligible, and intellection. To better understand the difference between divine and human thought, we will first distinguish the three terms: the *intelligible*, the *intellect*, and *intellection*. That which is *intelligible* are the forms or ideas of things. But to the extent that they inform things, ideas are only potential. They do not become active, or fully become ideas, until they are perceived by an intellect. The *intellect* is the ability to comprehend or perceive ideas, a characteristic of intelligent beings. As long as it perceives forms, the intellect remains a potential. It only becomes active when it identifies with the ideas that it embraces, by becoming one with them. *Intellection* is the becoming-idea of the intellect, the movement that realizes the union of the intellect and the intelligible. It is through intellection that the intelligent being passes from potentiality to act, and thus reunites with his essence. In accomplishing this act the intelligent being is united with his intellect and thus, at the same time, with the intelligible form comprised by the intellect. Each time there is an active thought, the intellect, the intelligent, and the intelligible become one and the same.

Why is man not always actively intelligent and thus united to his essence? Because mankind's intellect is not in a continual state of operation: It sleeps, dreams, grows tired, sick, etc. Even when it is "operational," man can choose not to employ his intellect by giving in to distraction, by allowing himself to be overcome by his passions, by immersing himself in the sensible rather than turning toward the contemplation of intelligible forms. Finally, assuming that a human being never sleeps, is never tired or sick, he would still be an active intelligence only intermittently. We do not refer to someone as intelligent if he always contemplates a *single* intelligible form. It is necessary that he proceed successively from the intellection of one idea to that of another. This inevitably results in discontinuities, breaks, which are exacerbated to the extent that human intellect receives its ideas from the external world. At the moment when our intellect leaps from one perception to another, intellection is interrupted and separated from its own essence.

God alone can contemplate a single idea since that idea is the source of all the others. This occurs without any sense of discontinuity because He *is* that idea. Being incorporeal, God never sleeps, never tires, is never immersed in sensation or passion. As pure idea from which all ideas emanate, He contemplates himself through the motionless movement of endless intellection. His essence is to be eternally intellect, intelligible, and intelligent.

While the thought of individuals is discontinuous because they sleep, grow ill, tired, or take vacations, the collective intellect is always alert. When a mind slips into sleep, a hundred others rise up to take its place. Consequently, the virtual world is always illuminated, animated by the flames of living intelligences. By combining thousands of intermittent flickering rays, we obtain a collective light that shines continuously.

It would be wrong to compare this process to some simple mental "changing of the guard," a summation of consciousnesses. No, this is the expression of a collective intelligence. For when I sleep, my angel continues to act in the virtual world. My angel. This is the expression I wish to give to my memory, my knowledge, my navigations, my desire to learn, my hierarchy of interests, the relationships I have with other members of the thinking community. This angel, my digital messenger, helps inform, orient, and continuously evaluate the virtual world, which is itself the expression of all messengers. Thus, when a member of the thinking community clothes his angelic body, he does not limit himself to a single flame in the darkness, he is immediately situated in a changing intellectual landscape, one that is diverse, crisscrossed with tensions, one that is formed by the virtual union of individual intelligences. He plunges into a space of communication, of calls and responses. He evolves within a universe of shared significations and common problems.

These landscapes will be mapped and divided into interactive spaces by new sign systems, dynamic diagrams and ideograms, moving architectures of images. Computer hardware and software will make these universes of signification sensible, explorable, and interactive.

Human intelligence? Its space is dispersion. Its time, the eclipse. Its knowledge, the fragment. Collective intelligence realizes its reintegration. It constructs transpersonal but continuous thought. An anonymous cogitation but one that is perpetually alive, uniformly irrigated, metamorphic. Through the intermediary of virtual worlds, we can not only exchange information but think together, share our memories and our plans to produce a cooperative brain.

Media communication has, of course, already established continuity in space and time: the telephone, facsimile, electronic mail, digital and telematic networks, radio, television, the press, etc. This does not, however, reflect a continuity of active and living thought that is everywhere singular and differentiated, emergent and convergent, but rather a network for transporting information. Are television viewers able to recognize one another? Do they combine their experiences and intellectual strengths? Do they collectively negotiate and perfect new mental models of a given situation? Do they exchange ideas? No. Their brains do not yet cooperate with one another. Media continuity is merely physical. It is a necessary but not sufficient condition for intellectual continuity.

The Sensible and the Intelligible

Until recently, writing was undoubtedly one of the most efficient means ever devised for producing collective thought. The network of libraries records the creation and experience of a multitude of human beings, dead and living. Reading and interpretation, from generation to generation, reestablish the fragile thread of memory, reactualizing dormant thoughts. Translation, from one language or discipline to another, enables disconnected thought spaces to communicate with one another. But conventional writing is by nature a system of static and discontinuous traces. It is an inert body, fragmented, dispersed, ever growing, whose consolidation and animation require of each individual a lengthy process of research, interpretation, and association.

To remedy this situation, the virtual worlds of collective intelligence will develop new forms of writing: animated pictograms, cinelanguages that will retain the trace of the interactions of their navigators. Collective memory will organize and redeploy itself for each navigator on the basis of his interests and travels in the virtual world. The new, angelic space of signs will be sensible, active, intelligent, and at the service of its explorers.

What is interpretation? The subtle mind attempting to coax the inert body of the letter into graceful motion. The evocation of an author's breath in the presence of dead signs. The dangerous reconstruction of the knot of affects and images from which the text arises. And finally, the production of a new text, the interpreter's text. But if the signs are alive? If the text-image or thought-space grows, proliferates, and metamorphoses continuously, following the rhythm of the collective intelligence? If the lead characters make room for the very substance of the angels? If the opaque and gigantic stratification of texts effaces itself before a fluid and continuous medium whose explorer always occupies the center?

From the confrontation of the living spirit and the dead letter, from the dialectic of the corpus and oral tradition, there arises the possible participation of all in the adventure of a people of signs in motion.

Within theological discourse the individual is not always active intelligence because he is often occupied with sensation rather than with intelligible forms. Yet the separation of the sensible and the intelligible is obviously not as clear as our neo-Platonic medieval philosophers assumed. As we have shown, every thought, even the most abstract, is based to some extent on the presence of an image. Within the domain of intellectual technology, progress consists in visualizing the very small or the very far, in making opaque matter diaphanous, in diagramming the inextricable complexity of processes, of creating images of abstract mental models, using maps to comprehend far-flung territories. In terms of an anthropology of the tools of thought, the intelligible is no more than a synthetic or diagrammatic version of the sensible.[10] If we can establish that the sensible and the intelligible are two poles of a continuum, the separation between man and active intelligence will not be as great as was previously assumed.

Of course the image is not used exclusively to improve understanding; it fascinates, seduces, and deceives as well. It is as if the sensible were the prize in a confused and inconclusive struggle: instrument of knowledge, playing field for the intelligence, or black hole of the mind? The more the intelligible is understood in terms of the sensible, the more the sign-image increases in scope and complexity, and the shorter becomes the distance between man and God. For theology, the divine influx enters the rational soul to nourish the imaginative soul. In the same way, the procession of celestial intelligences and souls began with pure intelligences, then descended toward the souls inhabited by spiritual imagination. Reversing the sense of direction, our model postulates that intelligence is first alerted by sensation, image, or, at least, by the sensible sign.

Stimulating the human mind, the new agent intellect is defined as a machine to make thought visible, to image abstraction and complexity, creating a landscape that our angelic bodies can explore, feel, and modify. The virtual world makes interconnected relations apparent, enables us to touch the most obscure notions; it illuminates images, makes them comprehensible. It is the most appropriate environment for the flowering and development of the visual[11] languages that will weave together the intelligence or, rather, the collective imagination.

Within the theological framework man receives his ideas from the outside, whereas God contemplates Himself. It is true that on a human scale, intelligence is the gateway to what is external, perpetual unfulfillment, an effort to reach the outside and that which is not self. When we learn, we enter the world of the other. But while learning, that is, while transforming himself, the thinking subject strives continuously to bring the stranger to him, to transform the other into self, so that foreignness can no longer be grasped in itself and we must again clear a path to the outside. Just as for Aristotle the soul was the form of the body, for us our intelligence is like the form or envelope of our world. The world that thinks in us. It is our responsibility to ensure that this envelope can unfold and grow to cover an increasingly vast and diverse world—or that it can filter and sort the figures it encounters to compose a more beautiful world—rather than harden, become opaque, and fold in upon itself. If our personal intelligence is the soul of a small world, collective intellects envelop much larger and more varied worlds. They enrich our thought to the extent that we participate in them, and their thought is improved as they incorporate other souls and other worlds. So long as its nomadic members continue to discover new dimensions and infuse it with fresh air from the outside, the collective intellect, by contemplating the virtual space that expresses its diversity, can think itself and the world it envelops.

The collective intellect thinks throughout time and space, perpetually releasing the thought of its members. For the thinking community to which we appeal through our prayers, as for the God of Avicenna or Maimonides, the intellect and the intelligible are one and the same. We refer to this union of the intellect and the intelligible in a collective being as its virtual world. It is at the same time a society of animated signs, a shared organ of perception, cooperative memory, and space for communication and navigation.

With respect to the intellection of the collective intellect, it resides, now and forever, in the experience, apprenticeship, and mental gestures of its individual members. It merges the journeys, negotiations, contacts, decisions, and effective actions of those involved in the continuous creation of a shared world. Only living and actual people can make collective intelligence active. For the virtual world is no more than a substrate for cognitive, social, and affective processes that take place among actual individuals. Just as writing or the telephone did not prevent people from continuing to meet in person, the virtual worlds of collective intellect cannot substitute for direct human contact. On the contrary, they should enable those who so desire to find one another and extend their personal, professional, political, or other relations. The virtual world is clearly the appropriate medium for collective intelligence, but it is neither its exclusive locus, nor source, nor goal.

The Three Freedoms

To better protect collective intelligences from the effects of alienation, we turn once again to theology. We know that God is His own cause. But what is a cause? According to Aristotelian philosophy, there are four types of cause: if, for example, we take a vase made by a potter, the clay is its *material cause*, the potter its *efficient cause*, the shape initially conceived by the potter its *formal cause*, and the property of containing liquid its *final cause*.

God obviously has no material cause. As for the rest, He is final, efficient, and formal cause in and of Himself, which is why He is absolutely free. A human being unfortunately does not have this possibility: His parents are his efficient cause, God (or biological evolution) his formal cause, and he can't always become an end for himself. But since naked and solitary man is unable to do so, why not attempt to form collective intellects capable of achieving such divine freedom?

The collective intellect is its own final cause. It has no goal other than to grow, develop, differentiate itself, and propagate the varieties of signs that populate it, the cosmic diversity that it envelops, and the ontological plurality that is its richness and its life. To do so, it must obviously maintain its existence and thus respect certain economic, technical, and other constraints.

The collective intellect is as much as possible its own efficient cause. It is born from the will of its members and not from some outside impulse. In a sense, therefore, it must already exist before it can come into being (since it is composed of "its members"). This paradox of creative circularity is inherent in all autonomic or autopoietic production. It is what makes the problem of the origin of life so difficult to resolve, torments the political philosopher (what is a society founded on if not a preexisting society?), and is the despair of educators (how can we enable a dependent being to grow accustomed to freedom?). It would be unrealistic to dissimulate the difficulties involved in initially forming collective intellects. Difficult, but not impossible. Life goes on, societies are formed, some beings manage to achieve a kind of freedom. The first steps will be small, modest, imperfect, before the dynamic of collective intelligence finally takes shape and grows.

The collective intellect is its own formal cause. Its appearance is not conferred upon it by some external entity. It emerges continuously from the multitude of free relations that are formed within it. Far from being represented by some discrete entity that oversees and structures it, collective intellect expresses itself within an immanent space. Freed of transcendent unity, it will continue to produce and reproduce the folds of its envelope, to redetermine that which will populate its world. It does not need to elect representatives or destroy idols for it to change shape or transform itself, since it is by means of a single continuous movement that it creates itself, knows itself, and produces its own image. The ability of the collective intellect to become its own formal cause is its greatest achievement, the touchstone of its immanence.

I would even say that its final cause is its own existence, that is, its existence as self-cause, in the sense that we have just defined. If its freedom is lost, it would be preferable to dissolve what remains of it, for it will no longer promote the freedom of those who constitute it.

But if its members succeed in maintaining the autonomy of the collective intellect, each increase in qualitative diversity strengthens the interest of everyone in its continued existence. And the more its members are involved in its permanent re-creation, the more the immanent dynamic of expression will favor the proliferation of ways of being. Each mode of freedom will be reflected upon the others in a positive spiral.

In this way collective intellect creates a new space.

6

The Art and Architecture of Cyberspace

The Aesthetics of Collective Intelligence

Cyberspace Under Construction

Communications networks and digital memories will soon incorporate nearly all forms of representation and messages in circulation. At this point I would like to consider the significance of the risks involved in developing these networks. Politics and aesthetics confront one another within the unbounded construction site of cyberspace. The perspective of collective intelligence is only one possible approach, however. Cyberspace might also presage, or even incarnate, the terrifying, often inhuman future revealed to us by science fiction: the cataloging of the individual, the processing of delocalized data, the anonymous exercise of power, implacable techno-financial empires, social implosion, the annihilation of memory, real-time warfare among maddened and out-of-control clones. Nevertheless, a virtual world of collective intelligence could just as easily be as replete with culture, beauty, intellect, and knowledge, as a Greek temple, a Gothic cathedral, a Florentine palazzo, the *Encyclopédie* of Diderot and d'Alembert, or the Constitution of the United States. A site that harbors unimagined language galaxies, enables unknown social temporalities to blossom, reinvents the social bond, perfects democracy, and forges unknown paths of knowledge among men. But to do so we must fully inhabit this site; it must be designated, recognized as a potential for beauty, thought, and new forms of social regulation. I would like to end this first part of the book with a discussion of the aesthetic dimension associated with engineering the social bond, which primarily focuses on the design of cyberspace and the use of creative play within this new environment of communication and thought.

Cyberspace. Of American origin, the word was used for the first time in 1984 by the science-fiction writer William Gibson in his novel *Neuromancer*.[1] Cyberspace designates the universe of digital networks as a world of interaction and adventure, the site of global conflicts, a new economic and cultural frontier. There currently exists in the world a wide array of literary, musical, artistic, even political cultures, all claiming the title of "cyberculture." But cyberspace refers less to the new media of information transmission than to original modes of creation and navigation within knowledge, and the social relations they bring about. These would include, in no particular order: hyptertext, the World Wide Web, interactive multimedia, video games, simulations, virtual reality, telepresence, augmented reality (whereby our physical environment is enhanced with networks of sensors and intelligent modules), groupware (for collaborative activities), neuro-mimetic programs, artificial life, expert systems, etc. All of these tools are combined in exploiting the molecular character of digitized information. Various hybrids of the above technologies and conventional media (telephone, film, television, books, newspapers, museums) will come into existence in the near future. Cyberspace constitutes a vast, unlimited field, still partially indeterminate, which shouldn't be reduced to only one of its many components. It is designed to interconnect and provide an interface for the various methods of creation, recording, communication, and simulation.

While the true "great works" remain to be accomplished within the universe of digital information and at the new sites for the emergence of collective intelligence, we continue to encumber the landscape with cement, glass, and steel. We have built pyramids when we are in the process of again becoming nomads, when an architecture for a new exodus is needed. In the silence of thought, we will travel the digital avenues of cyberspace, inhabit weightless mansions that will now constitute our subjectivity. Cyberspace: urban nomad, software engineering, the liquid architecture of the knowledge space. It brings with it methods of collective perception, feeling, remembrance, working, playing, and being. It is an interior architecture, an unfinished system of intelligence hardware, a gyrating city with its rooftops of signs. The development of cyberspace, the quintessential medium of communication and thought, is one of the principal aesthetic and political challenges of the coming century.

Digital interactive multimedia, for example, explicitly poses the question of the end of logocentrism, the destitution of the supremacy of discourse over other modes of communication. It is likely that human language appeared simultaneously in several forms: oral, gestural, musical, iconic, plastic, each individual expression activating a given region of a semiotic continuum, bouncing back and forth from one language to another, from one meaning to another, following the rhizomes of signification, increasing the powers of mind as it traversed the body and its affects. The systems of domination founded on writing have isolated language, established its mastery over a semiotic territory that has been cut up, parceled out, and judged in terms of a sovereign *logos*. The appearance of hypermedia, however, sketches an interesting possibility (among others that are less interesting): A resurgence that lies well within the path opened by writing falls short of a triumphant logocentrism and moves toward the rediscovery of a deterritorialized semiotic plane. But such a resurgence will be enriched with the powers of the text; it will be based on instruments that were unknown during the Paleolithic and are capable of bringing signs to life. Rather than limiting ourselves to the facile opposition between reasonable text and fascinating image, shouldn't we attempt to explore the richer, subtler, more refined possibilities of thought and expression created by virtual worlds, multimodal simulations, dynamic writing media?

In evaluating these new technologies, should we limit ourselves to the concepts of the information highway, telecommuting, interactive CDs, and virtual reality gaming that are presented for public consumption by the media? In doing so we lose sight of the continuity between these spectacular phenomena and the invisible, day-to-day use of existing intellectual technologies. When such new technologies are presented as unrelated phenomena, as objects fallen from the sky, we lose sight of the open and dynamic system they form, their interconnection in cyberspace, their contentious insertion in ongoing cultural processes. We remain blind to the different possibilities they offer to human becoming, possibilities whose full scope is rarely perceived and which should be the subject of deliberation, choice, and judgments of taste, rather than the fiefdom of technical specialists. Even in terms of the apparatus of communication and thought, we are neglecting the dimension of interiority and collective subjectivity, ethics, and sensibility that even the most seemingly technical decisions imply.

From Design to Implementation

With respect to its relationship to future projects, cyberspace will assume the form of a cultural attractor, which we can summarize as follows.

1. Called, controlled, dismissed, distanced, combined, etc., no matter how they are orchestrated, messages, regardless of type, will now revolve around the individual receiver (the opposite of the situation represented by the mass media).
2. The distinctions between authors and readers, producers and spectators, creators and interpreters will blend to form a reading-writing continuum, which will extend from machine and network designers to the ultimate recipient, each helping to sustain the activity of the others (dissolution of the signature).
3. The distinction between the message and the work of art, envisaged as a microterritory attributed to an author, is fading. Representation is now subject to sampling, mixing, and reuse. Depending on the emerging pragmatics of creation and communication, a nomadic distribution of information will fluctuate around an immense deterritorialized semiotic plane. It is therefore natural that creative effort be shifted from the message itself to the means, processes, languages, dynamic architectures, and environments used for its implementation.

Some of the questions that artists have been asking since the end of the nineteenth century will thus become more urgent with the emergence of cyberspace. These questions are directly concerned with the question of the frame: the limits of a work, its exhibition, reception, reproduction, distribution, interpretation, and the various forms of separation they imply. Under the present circumstances, however, no form of closure will be able to contain deterritorialization *in extremis*—a leap into a new space will be required. Mutation will occur in a sociotechnical environment in which works of art proliferate and are distributed. Yet, is it reasonable to even speak of a work of art in the context of cyberspace?

For the past several centuries in the West, artistic phenomena have been presented roughly as follows: a person (artist) signs an object or individual message (the work), which other persons (recipients, the public, critics) perceive, appreciate, read, interpret, evaluate. Regardless of the function of the work (religious, decorative, subversive, etc.) or its capacity to transcend function in search of the core of enigma and emotion that inhabits us, it is inscribed within a conventional pattern of communication. Transmitter and receiver are clearly differentiated and their roles uniquely assigned. The emerging technocultural environment, however, will encourage the development of new kinds of art, ignoring the separation between transmission and reception, composition and interpretation. Nevertheless, the ongoing mutation creates a realm of the possible that may never be realized or only incompletely. Our primary goal should be to prevent closure from occurring too quickly, before the possible has an opportunity to deploy the variety of its richness. With the disappearance of a traditional public, this new form of art will experiment with different modalities of communication and creation.

Rather than distribute a message to recipients who are outside the process of creation and invited to give meaning to a work of art belatedly, the artist now attempts to construct an environment, a system of communication and production, a collective event that implies its recipients, transforms interpreters into actors, enables interpretation to enter the loop with collective action. Clearly the "open work" prefigures such an arrangement. But it remains trapped in the hermeneutic paradigm. The recipients of the open work are invited to fill in the blanks, choose among possible meanings, confront the divergences among their interpretations. In all cases it involves the magnification and exploration of the possibilities of an unfinished monument, a succession of initials in a guest book signed by the artist. But the art of implication doesn't constitute a work of art at all, even one that is open or indefinite. It brings forth a process, attempts to open a career to autonomous lives, provides an introduction to the growth and habitation of a world. It places us within a creative cycle, a living environment of which we are always already the coauthors. Work in progress? The accent has now shifted from work to progress. Its embodiment is manifested in moments, places, collective dynamics, but no longer in individuals. It is an art without a signature.

The classic work of art is a gamble. The more it transmutes the language on which it rides, be it musical, plastic, verbal, or other, the more its author runs the risk of incomprehension and obscurity. But the larger the stake—the degree of change or fusion to which its language is subject—the greater the potential gain: the creation of an event in the history of a culture. Yet this game of language, this wager on incomprehension and recognition, is not restricted to artists alone. Each of us in our own way, as soon as we express ourselves, produces, reproduces, and alters language. From singular utterances to creative listening, languages emerge and drift along the stream of communication, borne by thousands of voices that call and respond to one another, take risks, provoke and deceive, hurling words, expressions, and new accents across the abyss of non-sense. In this way an artist can appropriate an expression inherited from earlier generations and help it evolve. This is one of the primary social functions of art: participation in the continuous invention of the languages and signs of a community. But the creator of language is always a community.

Radicalizing the classical function of the work of art, the art of implication creates tension and provides us with sign machines that will enable us to invent our languages. Critics may claim that we have been producing languages forever. True, but without our awareness. To avoid trembling in the face of our own audacity, to mask the void beneath our feet, or simply because this activity has been so slow that it has become invisible, or because it has had to encompass masses of people in constant motion, we have preferred the illusion of foundation. But the price of this illusion has been our sense of defeat. Powerless before the language of the absolute, overcome by the transcendence of the *logos*, exhausted in the presence of the artist's inspired effusions, castigated and corrected, bearing the weight of forgotten tongues, we falter beneath the exteriority of language. The art of implication, which can only give some idea of its true scope in cyberspace, by organizing cyberspace, is an art of therapy. It encourages us to experiment with the collective invention of a language that recognizes itself as such. And in so doing, it points to the very essence of artistic creation.

Having stepped out of the bath of life, far removed from their areas of competence, isolated from one another, individuals finally "have nothing to say." The difficulty lies in trying to comprehend them—in both the emotional and topological sense—as a group, in engaging them in an adventure in which they enjoy imagining, exploring, and constructing sentient environments together. Even if live and real-time technologies play a role in this undertaking, the time experienced by the imagining community overflows the staccato, accelerated, quasi-punctual temporality of "interactivity." The inadequacy of the immediate, of amnesiac channel hopping, no longer leads to lengthy sequences of interpretation, the infinite patience of tradition, which encompasses in a single sweep the ages of the living and the dead, and employs the quick currents of the present to erect a wall against time. In much the same way as madrepores erect coral reefs, commentary, strata upon strata, is always transformed into a subject for commentary. The rhythm of the imagining community resembles a very slow dance, a slow-motion choreography, in which gestures are slowly adjusted and respond with infinite precaution, in which the dancers gradually discover the secret *tempi* that will enable them to shift in and out of phase. Each learns from the others how to make their entrance in stately, slow, and complicated synchrony. Time in the intelligent community spreads itself out, blends with itself, and calmly gathers itself together like the constantly renewed outline of the delta of a great river. The imagining collective comes into being so that it may take the time to invent the ceremony by which it is introduced, which is at the same time a celebration of origin and origin itself, still undetermined.

Employing all the resources of cyberspace, the art of implication reveals the priority of music. But how can a symphony be created from the buzz of voices? Lacking a score, how can we progress from the murmur of the crowd to a chorus? The collective intellect continuously questions the social contract, maintains the group in a nascent state. Paradoxically, this requires time: the time to make sure people are involved, time to forge bonds, to bring objects into being, shared landscapes ... the time to return. From the point of view of a watch or a calendar, the temporality of the imagining collective might seem displaced in time, interrupted, fragmented. But everything occurs within the obscure, invisible folds of the collective itself: the melodic line, the emotional tonality, the hidden intervals, the correspondences, the continuity that it weaves within the hearts of the individuals who compose it.

For an Architecture of Deterritorialization

The artists who explore such alternatives may be the pathfinders of the new architecture of cyberspace, which will undoubtedly become one of the major arts of the twenty-first century. The new architects could just as easily be engineers, network or interface designers, software programmers, international standards organizations, information lawyers, etc., as individuals with a background in traditional forms of art. In this field the most obviously "technical" choices will have considerable political, economic, and cultural impact. We know that traditional architects and urban planners have helped produce our material, practical, and even symbolic environment. In the same way, the sponsors, designers, and engineers of cyberspace will help produce the environments of thought (sign systems, intellectual technologies), action (telecommuting, remote operation), and communication (access rights, rate policies) that will, to a large extent, structure social and cultural developments.

To guide the construction of cyberspace, to help us choose among the different possible orientations or even imagine new ones, some criteria of ethical and political selection are needed, an organizing vision. Means that contribute to the production of a collective intelligence or imagination should be encouraged. In keeping with this general principle, I would suggest that we concentrate on the following:

1. Instruments that promote the development of the social bond through apprenticeship and the exchange of knowledge
2. Methods of communication that are predisposed to acknowledge, integrate, and restore diversity rather than simply reproduce traditional media-driven forms of distribution
3. Systems that promote the emergence of autonomous beings, regardless of the nature of the system (pedagogical, artistic, etc.) or the beings involved (individuals, groups, works of art, artificial creatures)
4. Semiotic engineering that will enable us to exploit and enhance, for the benefit of the greatest number, the veins of data, the capital of skills, and symbolic power accumulated by humanity

In terms of the creation and management of signs, the transmission of knowledge, the development of living and thinking spaces, the best propaedeutic is obviously supplied by literature, art, philosophy, and high culture in general. Barbarism is born of separation. Contrary to what they may think, technicians have a great deal to learn from humanists in this area. Likewise, those in the humanities must make an effort to employ the new tools, since they redefine the work of intelligence and sensation. Lacking such interaction, we will ultimately produce nothing more than a meaningless technology and a dead culture.

I am making a case here for an architecture without foundations, similar to a boat, with its system of practical oceanography and navigation. Not some benign symbolic structure, analogous to a static image of the body or mind, the reflection of a stable world. On the contrary, the architecture of the exodus will give rise to a nomadic cosmos that travels the universe of expanding signs; it will bring about endless metamorphoses of bodies; within the fissure of flesh and time, it will dispatch its fleets toward the inviolate archipelagos of memory. Far from engendering a theater of representation, the architecture of the future will assemble rafts of icons to help us cross the seas of chaos. Attentive to the voice of the collective brain, translating the thought of plurality, it will erect sonorous palaces, cities of voices and chants, instantaneous, luminous, and dancing like flames.

II

The Knowledge Space

7

The Four Spaces

The Earth

The earth, the great nomadic earth, was the first space occupied by humanity. Our species secreted the earth while creating the world we inhabit. The earth is the world of signification that flowered during the Paleolithic period into our language, technology, and social institutions. Humanity invented itself by allowing the earth to unfold around it, a nourishing and responsive earth, an earth it perpetually re-creates through its chants and rituals.

The earth corresponds neither to the originating soil nor the period of our initial development. It is the immemorial space-time to which we cannot assign an origin. It is the space that is "always already there," that contains and overflows the beginning, unfolding, and future of the human world. The earth is not a planet, not even a biosphere, but a cosmos in which humanity communicates with animals, plants, landscapes, locales, and spirits. The earth is the space in which mankind, the stones, vegetables, beasts, and gods meet, talk, come together, and separate in a process of unending re-creation. The metamorphoses known to Ovid, Empedocles, and Lucretius, those that peopled the dreams of the Australian aborigines, those of the great mythic tales, all occurred on earth.

On earth, everything is real, everything is present, and the trance vision or god-inspired utterance do not necessarily contradict our sense of judgment. Upon the great nomadic earth, dream and awakening do not blend but support, interpret, and nourish one another. Animals live in an ecological niche. But humanity immediately exceeds any niche; humanity lives on an earth it constantly creates and re-creates through its languages, tools, and complicated but subtle social systems in which the cosmos is entwined. Mankind does not live within a niche, as a dog does, because it can gaze at the stars, invent the gods that fashioned it, adopt eagles and leopards as its ancestors, and live among its signs, its tales, and its dead. Man is the only animal that lives in the cosmos, that does not simply belong to a species but chooses its totems. Humanity is the species dedicated to the earth, to the cosmos of animals and plants in communication, to the *chaosmos* of metamorphoses.

The Neolithic revolution obviously did not eliminate the great savage and nomadic earth, the immemorial "garden." The *unconscious* is a poor term to express this permanence of the earth because it refers to a small, diminished, individual, and familial sphere from which the cosmos has been excluded. The great chaosmic earth is always present, resonating distantly beneath our feet, beneath the concrete, beneath the derisive signs of the spectacle. Penetrating the borders of our identities, the heart of the earth continues to sing its mad song of dream and life, the song that sustains the world's existence.

Territory

For the past twelve thousand years, a second anthropological space, the territorial space, has been spreading across the earth in ever expanding sheets, isolated patches that have slowly come together over the centuries. The domestication and rearing of animals, agriculture, the city and state, writing, the strict social division of labor—the order of appearance of such innovations has differed from place to place. But whenever they connect and mutually reinforce one another, they acquire an irreversible force, a power of expansion, a permanence, such that a new reality is established, the sedentary world of civilization.

The territorial space first came into being in the Near East, somewhere between the fertile crescent, Iran, and Anatolia. But there was also a Chinese Neolithic that appeared later on and, later still, a Mexican and Incan Neolithic. The exact dates aren't important. The Neolithic isn't really considered a historical period but an atemporal anthropological space, which, once it appeared, immediately reverberated throughout mankind's past and future. Agriculture, the city, state, and writing were from then on virtualities inherent in humanity, reflecting one another and contributing, each in its own way, to the partitioning of the territory.

Consider a Magdalenian hunter painting a large reindeer on the wall of a cave, or an Australian aborigine, naked, spear in hand, chanting the dream of his clan down a timeless path. They are typical emblems of the first space, or earth. On the stele that marks the boundary of the territory are engraved the exploits of Sargon of Agade, King of the Four Lands, the first emperor in history, who, through his conquests, unified the city-states of Mesopotamia. Near the center of the territory lies a pyramid, its shadow falling across the fellaheen, artisans, scribes, and soldiers, reaching to the borders of the land. It dominates the fields of barley and spelt, the maze of irrigation canals, the towns with their squares and streets, their courtyards, temples, statues, and walls, a pyramid that will carry the pharaoh's mummy into the future.

Territorial space attempts to cover the great nomadic earth, push it toward the edges of the world. It channels rivers, dries swamps, clears impassable forests—forests that have been burning continuously since the dawn of the Neolithic—builds bridges across rivers and ravines. The paved streets snaking their way along the ridges resonate with the footsteps of its legions. Armies, police, administrators, tax and tribute collectors help civilize mankind, construct the territory from within, build a social pyramid in the collective soul of a people. Like a stillborn infant, the entire hierarchy bears within it the mummy of the pharaoh.

With the earth, the territory establishes a predatory and destructive relationship; it dominates, confines, encloses, inscribes, and measures it. But the rivers overflow their beds, the forest continues to grow, looters from the desert sack the hoard of treasure, women and men leave their fields and homes behind. The earth always returns; it bursts into the very center of the territory. Ibn Khaldoun considered the struggle between nomadic and sedentary principles, the conflict of the spaces, to be the very dialectic of history.

Wheat and lentils, rice and soy, beans and corn, cattle and sheep characterize these constructed and controlled landscapes. During the Neolithic period, mankind inhabited a renewed cosmos, closely tied to the past, in which new gods presided. Cities provided him with anonymity, a new freedom. Writing gave him access to history. This increased power, however, fell not to the individual but to the great social machine, the state. Humanity multiplied along the river banks, deltas, and on the fertile plains.

For the past three or four thousand years, up until the Second World War, the majority of humanity were peasants who inhabited the territory throughout a long Neolithic period, which the toppling of empires, the movements of entire peoples and a handful of technical innovations have barely affected.

The Commodity Space

When did signs first begin to circulate more rapidly, laterally, escaping the overbearing hierarchy of caste? At the dawn of the Greek miracle, with the invention of money and the alphabet? During the Renaissance, which witnessed the rise of printing, the first industry and mass media, a time when European explorers were discovering the world's continents and creating the first global markets? Or did the commodity space come into being in the eighteenth century, amid the smoke of the industrial revolution? This was not the customary space of exchange or commerce, but a new world built from the incessant circulation of money in an ever tightening, ever quickening loop. Bills of exchange, promissory notes, term drafts, securities, currencies, interest rates, finance, speculation, calculation.

This floating world, dispersed and inconsistent, initially affected no more than the surface and margins of social life. But in the wake of an extraordinary historical conjunction, it began to bring together its scattered elements: money, banking and credit, the policing of populations no longer subject to despotic empires, capital and technology, extended markets, laborers torn from their fields, a collective revery or desire that was already beginning to escape territorial space, in search of another space, other velocities. This new world succeeded in enlarging itself, living off its own life. Crossing borders, upsetting territorial hierarchies, the dance of money brought in its wake an accelerated movement, a rising tide of objects, signs, and individuals. Steamboats, railroads, automobiles, roads, accidents, highways, junkyards, trucks, cargo, tankers, airplanes, subways, transport, market outlets, circulation, distribution, saturation, motionless speed.

The commodity space was smoothed out, maintained, and enlarged by a deterritorializing machine, which suddenly sprang into being, feeding on everything in its path. Just as King Midas transformed whatever he touched into gold, capitalism transmutes into merchandise everything it draws into its orbit. Wheat, skins, wool, cotton, fabric, sheets, clothing, sewing machines, chemical products, fertilizer, drugs, canned goods, frozen foods, refrigerators, washing machines, tobacco, detergents, diapers, soap, countless objects, an accumulation of things, stores, inventory, warehouses, catalogs, supermarkets, packaging, cartons, shop windows, consumption, waste, landfills.

Capitalism can only function through the territorial state, by carrying along with it, in the wake of science and technology, the flux of signs in the terrestrial cosmos, a redefined cosmos, one reinterpreted as resource, a reconstructed cosmos, reconstituted, redeployed by science and technology, televised, simulated: electrons and proteins, dams along the Amur and Yangtze—the technocosm. Iron and coal mines, firedamp, steam engines, work, textile industries, tractors, harvester combines, pesticides, furnaces, oil wells, refineries, work, gas refineries, asthma, bronchitis, power stations, nuclear reactors, electricity, work, cables, networks, lights, neon, machine tools, robots, work, strikes, concrete, glass, steel, plastic, offices, business centers, meetings, work, work, work, unemployment.

When the commodity space assumes its autonomy within the territory, it doesn't simply abolish the preceding spaces but subordinates them, organizes them in terms of its own objectives. The old, Neolithic territory is distended, hybridized, crisscrossed, cracked, disjointed, enveloped by the merchant technocosm. Capitalism is deterritorializing and, for the past three centuries, industry and commerce have been the principal engines driving the evolution of human societies. In the mid-eighteenth century there were approximately 750 million people on earth; by the year 2000 there will be 6 billion. Amid the ruins of the earth and the eroded monuments of territory, 6 billion people will inhabit the technocosm, with its speed, its incessant stream of amnesiac images. Cannons, shells, armor, explosives, destroyers, machine guns, mustard gas, Zyklon B, cruisers, aircraft carriers, atom bombs, phosphorus bombs, fragmentation bombs, blast bombs, helicopters, missiles, radar, war, and destruction.

Marx made economy the infrastructure of human societies and examination of the modes of production the key to historical analysis because in the nineteenth century commerce was the dominating space. But there is no economic base without capitalism, not before and perhaps not ever. Nevertheless, the great cybernetic machine of capital, with its extraordinary powers of contraction and expansion, its flexibility, its ability to insinuate itself everywhere, to constantly reproduce the relations of commerce, its epidemic virulence, seems invincible, inexhaustible. Capitalism is irreversible. It *is* economy and has made economy the permanent dimension of human existence. There will always be a commodity space, as there will always be an earth and a territorial space. Typewriters, printing presses, newspapers, magazines, photographs, posters, publicity, cinema, stars, telephone, radio, easy listening, television, records, classical music, tape recorders, cassettes, hi-fi systems, rock music, baroque music, Walkmen, video games, interactive multimedia, world music, museums, rockets, satellites, computers, telematics, information, communication, databanks, organized travel, intelligent buildings, water and gas on every floor, R&D, genetic engineering, art, culture, spectacle.

Is there any anthropological dimension that will enable us to escape the vortex of capital? Are there any forces—more rapid, more enveloping than economic forces—that will deterritorialize deterritorialization? Weaving back and forth in its dizzying ascension, the mantle of commodity, the sky of contemporary humanity, tears open to reveal another space.

The Knowledge Space

The knowledge space doesn't exist. Etymologically, it is a u-topia, no-place. It is embodied nowhere. But if it is not realized, it is already virtual, waiting to be born. Or rather, it is already present, but buried, dispersed, travestied, intermingled, sprouting rhizomes here and there. It emerges in patches, traces, just below the surface. It flickers even before it has had a chance to develop its own autonomy, its irreversibility. This crystallization of a free knowledge space, the creation of a new anthropological dimension, the passage from a point of no return, may never take place.

Today, within the commodity space, the knowledge space is still subject to capital's need of competition and calculation. Within the territorial space, it is subordinate to the objectives of power and the bureaucratic management of the state. And on earth, it is caught up in the enclosed worlds and archaic mythologies of the new age or deep ecology, as if life's diversity, the stars, cosmic energy, the powerful images of the collective unconscious weren't also continuously reproduced by the dreams, signs, and machinations of mankind, as if there was but one nature, as if the great earth was not extravagant, plural, nomadic.

What is knowledge? Obviously it is not simply scientific knowledge, something recent, rare, and limited, but a form of knowledge that qualifies our species, *Homo sapiens*. Each time a human being organizes or reorganizes his relationship to himself or his peers, to things, signs, or the cosmos, he is engaged in a form of knowledge, apprenticeship. Knowledge, in the sense I am using the term, is a knowledge-of-living, a living-in-knowledge, one that is coextensive with life. It is part of a cosmopolitan and borderless space of relations and qualities, a space for the metamorphosis of relationships and the emergence of ways of being, a space in which the processes of individual and collective subjectivization come together.

Thought can't be reduced to so-called rational discourse. There are body-thoughts, affect-thoughts, percept-thoughts, sign-thoughts, concept-thoughts, gestural-thoughts, machine-thoughts, world-thoughts. The knowledge space is the plane of composition, recomposition, communication, and singularization, where thought triggers thought in a continual process. Site of the dissolution of separation, the knowledge space is inhabited, animated by collective intellects—imagining collectives—that are continuously being dynamically reconfigured. Collective intellects invent mutating languages, construct virtual universes, cyberspaces, in which unknown forms of communication seek one another. Yet this fourth space does not exist. It has not yet become autonomous. But in another sense, following its virtual emergence, its quality of being will be such that its cry will be heard throughout eternity: The knowledge space has always existed.

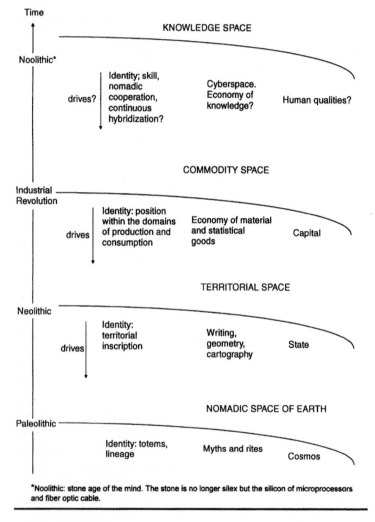

THE FOUR SPACES: EARTH, TERRITORY, COMMODITY, KNOWLEDGE

The knowledge space rejoins the earth, or rather, another earth, no longer immemorial, centered, and closed, but a sphere of artifice shot through with streaks of light and mutating signs, a cognitive planet traveling at breathtaking speed, an electronic storm that doubles, pluralizes, and deregulates the old nomadic earth of animals, plants, and gods. The knowledge space is not a return to the earth, but a return of the earth to itself, an overflight of the earth by itself at the speed of light, an uncontrolled cosmic diversification.

In light of the disturbing statistics that highlight the more troubling aspects of the earth, the territory, and the commodity universe, the coming of the third millennium harbors the seeds, the virtual figure of an autonomous knowledge space. Not the autonomy of scientific knowledge, which has been acquired over the past twenty-five hundred years, but a space of living-in-knowledge and collective thought that may be able to organize the existence and sociability of human communities. This fourth anthropological space, should it come to fruition, will harbor forms of self-organization and sociability that tend toward the production of subjectivity. Collective intellects will wander in search of unknown qualities and modalities of being. Not exactly an earthly paradise, since the other spaces, with their limitations, will continue to exist. The intention of collective intellect is not to destroy the earth, or the territory, or the market economy. On the contrary, the long-term existence of the first three spaces— excluding some merely barbaric and precarious form of survival—is without doubt conditioned by the resurgence of a new plane of existence for humanity.

There will be no magnificent twilight to usher in the knowledge space, only numerous pale dawns.

8

Anthropological Space

The Multiple Spaces of Signification

Our interactions produce, transform, and continuously develop heterogeneous and interlinked spaces. A simple conversation could be seen as the shared construction of a virtual space of signification, which each speaker attempts to shape according to his mood and intentions. These plastic spaces, which arise through the interaction among individuals, comprise messages, the representations they evoke, the speakers who exchange them, and the overall situation as it is produced and reproduced by the actions of the participants.

Lived spaces are relativistic. They bend and shape themselves around objects they contain and which organize them. The extent to which people, images, words, and concepts are capable of structuring a space of signification depends on the affective force associated with them. Evanescent spaces, like small bubbles that are formed when two people meet and then disappear...

more durable spaces, which are reused, enlarged, hardened, institutionalized.

Each day we experiment with such lived spaces as these, which are born of the interactions among individuals. But there are much larger, institutional spaces, formed by social groups, large cultural collectivities, which involve not only human beings but nonhuman elements of various origins: sign systems, means of communication, weapons, tools, electrons, viruses, molecules, etc.

We recognize the importance of an *event* within the intellectual, technical, social, or historical sphere by its ability to reorganize the proximities and distances in a given space, its ability to create new space-time(s), new proximity systems. Like the interpersonal spaces between us, cosmopolitan worlds intersect, dwindle, and transform themselves, displace their affective energies, leave in their wake of becoming new figures of desire.

Human beings do not inhabit only a physical or geometric space, they simultaneously live in emotional, aesthetic, social, and historical spaces, spaces of signification[1] in general.

The neighbor across the hall, with whom I exchange hellos and goodbyes, is situated quite near me in ordinary space-time. But while reading the work of an author who has been dead for three hundred years, I can establish, within the space of signs and thought, a much stronger intellectual connection than I have with my neighbor. The people standing around me on the subway are more distant, within affective space, than my daughter or father, who are three hundred miles away.

We live in thousands of different spaces, each with its own system of proximity (temporal, emotional, linguistic, etc.), such that a given entity can be near us in one space, yet quite distant in another. Each space has its own axiology, its own system of values or measurement. An object that is heavy within a given space will be light or marginal in another. A good part of our cognitive activity consists in helping us find our way among the multitude of different worlds in which we navigate. We must quickly determine the topography and axiology of the new spaces in which we participate, distinguish systems of values, understand and interpret the ways in which situations develop.

We spend our time modifying and improving the spaces in which we live, in connecting and separating them, articulating and solidifying them, introducing new objects, displacing the forces that structure them, jumping from one space to another.

Anthropological Spaces Are Structuring, Living, Autonomous, and Irreversible

Disparate spaces are organized around material or ideal mechanisms. They expand across a planet or proliferate within molecular environments. Some of them are slow, rigid, viscous; others come into being and quickly vanish. Among this multitude, what are the specific characteristics of the four anthropological spaces?

These spaces extend to the whole of humanity. They are themselves woven together from a bundle of interdependent spaces. Earth, territory, the commodity and knowledge spaces, are engendered by the practical and imaginative activity of millions of beings, by anthropological machines that work within the recesses of their subjects, within the depths of technological operations, cutting through the organizational structure of institutions.

The four anthropological spaces are structuring. They themselves contain or organize a large number of different spaces. These anthropological spaces are not to be understood as strata, abstract dimensions of analysis, the result of an analytical or purely chronological point of view. For the earth, the territory, the commodity and knowledge spaces are living worlds continuously engendered by the processes and interactions that unfold within them. Anthropological spaces push from within.

The knowledge space, for example, isn't simply the subject of cognitive science. The cognitive stratum is obviously present in all human activity. Man has been thinking since the origin of the species, but each anthropological space develops its own specific forms of knowledge. The knowledge space didn't begin to shape itself with any degree of consistency until the twentieth century. As an ongoing anthropological creation, it is a living plane, qualitatively differentiated, deployed by means of the metamorphoses and navigations of the collective intellects that pass through it. It is not some kind of abstract container for all possible knowledge. On the contrary, it harbors a specific kind of knowledge and reorganizes, hierarchizes, and submerges within the active environment it comprises, modes of knowledge from the other anthropological spaces.

Similarly, the commodity space is not economy, the subject of a specific social science. It is obvious that production and exchange have existed since time immemorial. Yet the world of signification, social relations, and interaction with the cosmos that unfolded after the industrial revolution, and continues to expand and proliferate, came into being at a specific moment. It has overflowed the domain of production and economic exchange to encompass nearly all aspects of human life. The commodity space is not simply a stratum of social life that has been dissected according to the methods of a specific science; it is a world that has grown and developed autonomously, a huge cosmopolitan machine, self-organized, creative, destructive.

Anthropological spaces appeared gradually along with the development of humanity. They assumed a certain consistency and organized themselves, and at some point became irreversible. The earth, as I have tried to show, cannot be separated from humanity as such. The disappearance of agriculture, the state, and writing is obviously not unthinkable, but its advent could only be seen as a terrifying catastrophe, a deadly chaos. The same would be true if capitalism were to completely collapse (rather than suffer an economic crisis or recession). It is their irreversibility that enables us to qualify these spaces as anthropological.

Anthropological Spaces Are Planes of Existence, Contingent and Eternal Velocities

Although they appear in succession and are super-imposed on one another, anthropological spaces are neither infrastructures nor superstructures that are mechanically determined or dialectically integrated. Each anthropological space harbors its own infrastructure. This infrastructure provides its autonomy and consistency, serves as its culmination in a way, and does not simply precede or determine it. Languages and narratives on earth, fields and clay tablets for the territory, printed matter and machines in the commodity space, digital networks, virtual universes, and artificial life in the knowledge space.

Anthropological spaces in themselves are neither infrastructures nor superstructures but planes of existence, frequencies, velocities, determined within the social spectrum. Suddenly, humanity finds itself moving at a greater speed. And this new speed brings a new space into being.

The earth provides the fundamental frequency. The first space corresponds to the introduction of velocities exceeding those of animal life: the velocity of language, technology, and culture. The territorial space introduces the first velocity perceptible at the individual level, that of writing and empire, bureaucracy and borders: torpor, the extended time of the territory. With capitalism comes acceleration. The knowledge space itself develops within the limits of real time, on the other side of "live." These four velocities, these four frequencies, coexist.

The appearance of the anthropological spaces is in no way governed by necessity. Evolution could just as easily have stopped with the large primates or turned out quite differently than it has. In which case, neither human speech nor the kernel of silence and enigma that inspires it would ever have made their appearance on earth. Humanity could have stabilized during the Paleolithic, as witnessed by the Australian aborigines and the many other hunters, gatherers, and nomads whose social and intellectual sophistication remains the equal of post-Neolithic cultures.

There is no compelling reason why civilization should ultimately have turned to capitalism. Mesopotamia, Egypt, Greece, Rome, China, Islam—the great empires could have succeeded one another, coexisted indefinitely, without the great movement of historical acceleration and economic unification of the world that began in the sixteenth century ever taking place. And although the knowledge space appears along the horizon of thousands of projects, thousands of currents of contemporary society, although its formation is in my opinion eminently desirable, it may never achieve autonomy. Thus the anthropological spaces are contingent. And yet, once they take shape, even virtually, they become eternal, timeless, as if they had always already been there.

The irreversibility of anthropological space affects the past as well. Don't our myths speak of a language that preceded the earth? According to Pierre Clastres,[2] primitive societies were organized around a political system that was "anti-state," a state they certainly had never encountered, but a conceivable, virtual, and threatening state. Isn't it equally possible, as Deleuze and Guattari[3] suggest, that borderless, "cosmopolitan" capitalism, in the sense of free enterprise, deterritorialization, and the general and unbridled acceleration of circulation, has been since its origin the state's secret nightmare? Is economy as a discipline anything more than the flattened, analytic form of the eternity of capital?

As mentioned previously, it would be a serious misconception to assimilate the anthropological spaces to a point of view, an analytical dissection of a preexisting reality, when such spaces are created and intersect from within. To assume that anthropological spaces are classes or sets in which we can place beings, signs, things, places, or individual entities of the human world would be an equally wrongheaded approach.

Anthropological spaces are worlds of signification and not rarefied categories sharing physical objects: a given phenomenon can thus occur in several spaces at once. Each will possess a different appearance, weight, and velocity. In place of a taxonomy, the instrument of orientation and identification used to organize the four spaces is a kind of anthropological *Carte du Tendre*. Love is not strictly limited to the summits of passion, the torrents of jealousy, or the swamps of boredom. It passes successively from one to the other, is present in several places at once. Similarly, the map of the anthropological spaces serves less to classify people, institutions, or events at a given location than to deploy, for each phenomenon, the full range of the anthropological spectrum.

Humanity crosses the four spaces with its entire being: walking, its feet strike the great earth of myth, its hair rising to the cosmos and the gods; sitting, composed, it is inscribed on the territory; its arms work the commodity space, eyes and ears devouring the signs of the spectacle; its head represents the knowledge space, a brain connected to other brains, secreting the virtual worlds of collective intellect, wandering, navigating, re-creating a thousand other earths upon the pluralized sphere of artifice.

It is less important to classify or arrange elements, than to identify, within a given institution, a given cosmopolitan machine, a given event or experience, that which is part of the cosmic, territorial, or merchant spaces, and that which clears the way to a fourth space, virtual and utopian, one which traces the lines of the future. Anthropological cartography is a checklist, a substrate of memory, a tool to help deploy every dimension of a being or process. If such an approach serves only to separate, classify, or isolate, we should abandon it at once.

We should also be cautious in accepting the uneventful succession or superposition suggested by an inevitably linear discourse. To gain some idea of the complexity involved in such an undertaking, imagine a notebook containing four pages (each page corresponding to an anthropological space) that have been torn, crumpled, and rolled into a ball. Now suppose that a needle (representing the phenomenon to be mapped according to our projection system) is stuck into this ball of paper. The needle will pierce the same space several times. Each new needle stuck in the ball will intersect the four spaces differently, both in terms of the order and the number of times it enters them.

Shredded, torn, crumpled, pierced with holes, inextricably folded within one another, earth, the territory, capital, and the virtual space of knowledge coexist everywhere differently.

9

Identity

Each space corresponds to a separate identity, a style of desire, a physical structure. There are terrestrial and territorial affects, commodity and knowledge affects.

Microcosm, Micropolis, and the Small House

On earth, names, tattoos, blazons, totems, and masks are all signs that signify identity. The definition of an individual by his participation in the clan, the blood line, the very ancient system of descendance and alliance, dates from the period of the great prehistoric earth. In its terrestrial manifestation, lineage cannot be encompassed by a narrowly familial definition, for it is cosmic as well. The ascendants of a human being are mythical ancestors, heroes, gods, animals, planets, a whole range of totemic, archetypal, or fundamental entities.

The being is formed by a network of cosmic relations that define him and determine his status. Exteriority becomes interiority: situated in the universe, the individual is a microcosm, an echo, a reflection of the whole. Each part of his body, each movement of his soul, reflects events or places that exist in the world.

The Neolithic revolution provided a means by which individuals were attached to the soil and their existence recorded by the state, a mechanism that exists to this day. The political territory has covered the earth of our ancestors. Its inhabitants have become sedentary. The state has turned lineage into an element of its own territorial strategies. Our relationship to territory, notably by means of landed property and its multiple avatars, defines the place of the individual in society, his identity. What rights do individuals have over a given portion of the land as property owners, suzerains, vassals, governors, farmers, sharecroppers, serfs, slaves? To what part of the land survey are they assigned? Do their names appear on some register of orthodoxy, some great book of civil government? Are they recorded on a tax roll? Though it might appear absurd on the great nomadic earth, we can ask what their address will be in this new space. Territorial identity is constructed around the house, the domain, the city, the province, the country. Man becomes a sedentary inhabitant.

But territorial identity is not limited to geography alone. It involves position and rank in institutions, castes, hierarchies, civil service corps (teachers, mining engineers), orders (doctors, architects, the nobility or clergy), disciplines (paleontology, sociology), everything that organizes a space in terms of borders, ranks, and levels. The identity provided by a diploma, for example, combining both rank and discipline, is obviously a component of the territory and not the knowledge space. Aside from a title of ownership and geographical registration, signs of territorial identity serve as insignia, grades, ranks, medals, and marks of affiliation of all sorts.

On earth man is a microcosm. Within territorial space on the other hand, the body is a hierarchized organism and the soul appears as a micropolis, a *micro polis*, a small state shaken by rebellion and passion, and acting against the empire of reason or law. The psychology of the territory is an internalized politics, just as its religion is the hypostasized image of the social order.

The commodity space deterritorializes and destructures the older frameworks of sociability and identity. Individuals are redefined according to their role in the fabrication, circulation, and consumption of goods, information, and images. Within the commodity space the signs of identity are quantified: income, salary, bank accounts, "external signs of wealth." We could say, to use an outdated terminology, that identity depends on where we fall within the system of production and our position within the circuits of consumption and exchange. As has been pointed out by a large body of sociological literature and fully exploited by advertising, consumption plays a greater role in constructing identity than the satisfaction of predetermined "needs." We are beginning to discover that the same is true of work. A curse within territorial space (and the territory is always present), labor tends to become a vector of socialization and self-affirmation within contemporary commodity space.

In a world dominated by economy, the individual is no longer a microcosm or a micropolis, but a *micro oïkos*. The word "economy" is derived, by way of Latin, from the Greek *oïkos*, house, and *nomia*, administration. Economy is the management of the household, proper organization of the home. Serving as a transmission channel for material goods through inheritance, the family also constitutes the individual's interiority. The soul is a theater where family tragedies are continuously enacted. Identity is constructed through the relationship between parents and children, by means of an Oedipal triangulation. From psychoanalysis to the bourgeois novel, the root of identity is located in childhood, within the shelter of the home. The family is no longer the clan, or some mythic or cosmic lineage, but the limited nucleus of parenthood. Psychology is no longer a politics but an economy of affects, the management of energy and libido.

Thus the capitalist machine deterritorializes and accelerates a number of social processes, tirelessly constructs new cosmopolitan mechanisms, while it paradoxically limits the extent of subjective identity, which, in the commodity space, gravitates around the family, work, and money.[1]

Toward a Sapient Identity, or Polycosmos

The emergence of a reality organized around knowledge can provoke a profound crisis of identity in which the older principles of self-orientation and identification in terms of a community lose their effectiveness. While nearly everyone has a name (alliance and lineage), is inscribed within the territorial space (at least by means of an address), is subject to the state, and participates to some extent in commodity production and consumption, the majority of individuals have no means of orienting themselves within the knowledge space. What concepts, what methods are needed to reveal the knowledge space and, at the same time, our individual identity within it? One possible direction is suggested by the *cosmopedia*[2] and the use of *knowledge trees*.[3] I will summarize some of the ideas developed in working with Michel Authier on these questions. Collective intellects emerge, interconnect, move, and change. It is through the circulation, association, and metamorphosis of thinking communities that the knowledge space is born and perpetuated. Each collective intellect harbors a virtual world that expresses the relations it maintains, the problems that activate it, the images it shapes from its environment, its memory, its knowledge in general. Members of the collective intellect coproduce, develop, and continuously modify the virtual world that expresses their community: the collective intellect is always learning, always inventing.

Within the knowledge space, individual identity is organized around dynamic images, images it produces through the exploration and transformation of the virtual realities in which it participates. The virtual body of the individual can be seen as the animated image of the hero or his vehicle (spaceship, race car, etc.) in a video game. The image that represents the player moves within a virtual world, combats his enemies, wins or loses lives, reaches goals, transforms his appearance or size, modifies his powers. But this analogy can only take us so far. A video game is determined by fixed rules, whereas a collective intellect is constantly questioning the laws of its immanent cosmos. A video game is first conceived by its designer, but the members of a collective intellect are both the designers of their cosmos and the heroes of the adventures that occur there: There is no longer any clear separation between the exploration and construction of the virtual world. The video game simulates a physical universe, whereas a collective intellect projects a space of signification and knowledge. A video game plunges the player into an imaginary territory, while the virtual world of collective intellect serves as a map, an instrument of location and orientation that refers to a real space, the most intensely real space we currently have, the space of living knowledge. The collective intellect recursively and cooperatively builds a cinemap of its world of signification, a hypermap linked to a multitude of thinking beings, works, and communities, a compass of the mind that points to other maps and other worlds.

The collective intellect constructs and reconstructs its identity through the intermediary of a virtual world it expresses. The individual, in turn, possesses many identities in the knowledge space, one for every virtual body he secretes in the cinemaps and cosmos of signification that he explores and helps create. Since he participates in more than one collective intellect, he creates several knowledge blazons.[4] Until the widespread acceptance of digital cosmopedias and knowledge trees that will make the knowledge space irreversible, we will have to settle for the experience of thinking. It's as if we were already located on the dynamic maps of shared knowledge, as if we were already navigating this fourth space, among the disturbances and currents of collective intelligence. On earth man is a *micro cosmos*, within territorial space, he is a *micro polis*, within commodity space, he has become a *micro oïkos*, a small house, and within the knowledge space, humanity is even more restricted: He is nothing more than a brain. Even his body becomes a cognitive system.[5] But the brain shapes itself collectively, makes contact with other brains, with systems of signs, language, and intellectual technologies,[6] it participates in thinking communities that explore and create multiple worlds. Thus the brain of *Homo sapiens sapiens* turns in upon itself, unveils its obverse and transforms itself into a *polycosm*. Within the knowledge space, humanity becomes nomadic once again, pluralizes its identity, explores heterogeneous worlds, is itself heterogeneous and multiple, in the process of becoming, thinking.

Quantum Identities

Before life, matter, or information could be processed at a sufficiently fine level of granularity, in each case an original microstructure landscape had to be created. Some means was needed to access, not some metaphysical atom, a final element, but a practical unit, a de facto element capable of providing a purchase on molecular technologies. But the gene of molecular biology, the byte of information, and the atom of nanotechnology are not trivial inventions. These grains are not the fragments of things, simple analytical residues, but the droplets of an ocean, vectors of a space in flux, signs of a remote language, the moving and pulverized backdrop out of which emerge—slow-motion vortex or momentarily frozen volute—the forms of materials, organisms, and messages. The practical units I am referring to are not, therefore, merely parts of complex systems. Their movements and metamorphoses smooth space-time continuums that are infinitely slower (genes) or faster (bytes, atoms) than the macro-entities that depend on them and adapt them in turn. Genes, atoms, and bytes experience their own adventures. Their interactions and paths describe worlds, interlacing histories that are distinct from those of animals, alloys, or texts. Macro-entities can't be reduced to micro-units. On the contrary, they offer the possibility of telling other stories, an additional degree of freedom.

From what grains will the images of free beings and the creators of meaning be formed? Is there a quantum mechanics of freedom? Can we measure subjectivity? We now know that these quanta of subjectivity will never be functional modules or parts of a system (no matter how complex), for this would mean the reification and, ultimately, annihilation of the subjectivities we claim to hold in such high esteem. The quanta of human qualities will be signs, and nothing more than signs. Will they lend themselves to exteriorized descriptions of their subjects, from the vantage point of some science of identity? No. The quantum mechanics of qualities can only be a collective mechanism of utterance, the space of a pluralistic speech, an instrument of self-description intersected by the individuals, groups, and situations from which they emerge. Such a language disdains images of the subject that assume the form of linear, hierarchic, or systemic structures, stratified into well-behaved and interlocking levels of integration. We should think of such subjects as chaotic concentrations of quanta, hot spots capable of generating new signs. Behold mankind: a moving cloud putting forth pseudopods, expanding, growing, adapting itself to a particular moment and terrain, to the flexible geometry of a limitless virtual plane. Quanta of subjectivity are like footsteps whose imprint delineates the trembling and incomplete image of a dance. They are responses whose risky, improvised interaction helps define a role that is even more evanescent.

A quantum of quality will always be proportional to a completed or potential act, some real or possible event. The purpose of human engineering is to enhance the most inconsequential of mankind's actions, promote its negotiation and evaluation, make it visible (the act, not necessarily the individual who accomplishes that act). Can we ensure that our actions will not be in vain? Each act is a potential speck of gold. The raw material of the economy of qualities is composed of human acts and potentialities. The products of such an industry are also human acts and potentialities, no longer raw, dispersed, forgotten, or neglected, but remembered, enhanced, oriented toward situations in which they will make sense, coordinated within a dynamic of power.

The quantum nature of quality is not based on conventional analytic and Cartesian rationality, for the quanta involved are emitted by subjects, they are semantic, not objective, solid, and fixed. These atoms of meaning, these grains of liberty, vary according to the context, as do all genuine signs. The word "horse" does not have the same value in the expression "get off your high horse" as it does when we say that "the mass of the solar system is greater than that of a horse." Semantic units are elastic and deformable, depending on the situations in which they are used and which they help construct. Like words, the quanta of quality have an independent existence (since we can compile dictionaries) but lack any definite meaning or value without a context. The step within the dance, the dance within the ballet. The response of an actor playing a role, and the role within the play. The sign of quality as element of the mask and the mask itself, sometimes dark, sometimes gleaming beneath the lights, randomly scattered by the movements of the farandole, shattered into a thousand pieces in the midst of the ball ... subjects no longer appear as solid figurines placed within clearly demarcated territories, but as nomadic distributions streaming through a moving space.

Closely linked to territorial space, identities of membership appear as an indistinguishable mass. They fix, separate, or blanket individual human qualities. The identity of membership is often characterized by an expression such as "I am an..." followed by a categorial name, or identity marker. It can be based on roots, origins, inclusion in a geographical, political, or functional (skill) group, some biological characteristic (age, sex, etc.). Regardless of its specific characterization, this type of identification nearly always culminates in the terrifying, lethal, and molar distinction between "us" and "them." The quantum nature of quality avoids such binary distinctions. By engineering the social bond, we allow another kind of subjectivity to develop, one that pulverizes the signs of knowledge or identity, enabling them to flow, blend, and run together, grow strong, expand, and intermingle. It does not shatter identities but liberates them: It provides everyone with the power to forge their own images. The quantum nature of quality doesn't enclose us in analytic definitions; it affords everyone a space for self-description. Far from being made transparent for the sake of power, the subject, from his unrepresentable core, becomes a participant. He decides whether or not to vend a given quality he possesses on a market. The process of

engineering the social bond does not manipulate subjectivities, it makes available to them the collective equipment that will facilitate the orientation, expression, and
evaluation of the self. It provides them with the tools of
subjectivization, which leaves open to them the possibility of appropriating their speech and controlling their
image.

Technically, the individual will be able to express
himself by distributing dynamic ideograms throughout
an indefinite number of virtual worlds. The economy of
human qualities must provide an alternative to subjectivization by inclusion. It can do this by allowing individuals
to freely project an unlimited number of images of themselves throughout an unrestricted variety of collective
spaces. Thus each one of us will be able to invent our
identities (our dances, our roles) by sharing in the construction of a large number of communities (balls, plays).
The individual becomes a molecular vector of collective
intelligence, multiplying his active surfaces, complicating
his interfaces, circulating among different communities,
simultaneously enriching both his own identity and
theirs.

Membership always involves a lessening of our
powers of being. Once free, the subjective cloud rises
from the past, folding and unfolding itself, now and toward its futures. Turning in upon itself, it releases to the
world, as an affirmation of self, the shifting diversity of
its rays.

Coexistence of the Four Identities

Each of us possesses four types of identity, even
though the first may be forgotten, even though the last
has yet to make its appearance. We are born and grow up
within all four spaces simultaneously. Our birth is not
simply territorial (an entry in a list of records) and famil-

ial. We are also born by and for the earth, and through our birth we inaugurate a cosmic existence. Succeeding spaces have obscured this ever active stratum of our identity, which rises to the surface in an illusory and dogmatic fashion in the form of astrology, spiritualism, esotericism, energy cults, and various new age movements. Astrology in particular offers a psychology, a system of signs to inform people and situations of an existence that escapes any political, economic, or familial determination, and provides the individual with a new relationship to the cosmos. This helps explain its continued success, even though the rational image of the universe that preceded it is no longer considered valid.

Those who claim they can heal humanity, repair identities, and supply us with ideals run the risk of focusing exclusively on the familial and commodity space, the territorial space. Obviously, social ills such as dysfunctional families, unemployment, and poverty need to be eradicated or at least treated. At times national pride offers some consolation. Our identification with institutions, power, hierarchies, and laws provides us with reassuring frameworks, "career outlooks." But it is clear that the solution to a number of our contemporary psychological, social, and cultural problems will be found in the discovery or rediscovery of other spaces. We must learn to displace our identities, our affects, our vital forces toward the earth, rediscover our relationship to the cosmos. As for the effective constitution of the knowledge space, it will provide us with a new space of freedom accessible to both communities and individuals. As of today, knowledge, thought, invention, and collective apprenticeship provide each of us with the ability to participate in a multiplicity of worlds, erect bridges across the separations, borders, and graduated levels of territorial space. Culminating in a plurality of universes of signification, the knowledge space may help us rediscover the earth. An identity made whole will cross the four spaces.

10

Semiotics

Semiotic of the Earth: Presence

Each anthropological space deploys its own regime of signs, a specific semiotic. On earth the sign participates in being and being in the sign. Everything speaks to us. Every event becomes a message and every person a messenger. Our least perception becomes an index, image, or symbol. Animals and men, stars and climates, shapes and details become signs for us, referring us to stories, utterances, rituals. "Perfumes, colors, and sounds call to one another"[1] along lines of affect, in accordance with the contiguities, analogies, and correspondences that organize the cosmos.

The sign is also an attribute, an active component of the thing, the being, or situation that it qualifies. Statues and masks give off mysterious forces. Words have power. Names radiate energy, qualities. Swept along by the breath on which it rides, the sign is never separated from a *presence*. Words become acts. They exercise power, de-

163

stroy, create. Images and fetishes act remotely. Divine acts and human rituals are the gestures and songs that support the world.

On earth the universe of signification reflects the reign of power and presence. Along the uniform space of the great nomadic earth, beings, signs, and things interconnect as rhizomes, exchange places, weave an unbroken canvas of meaning. This is the semiotic environment of primitives, animists, preliterate cultures, and very young children, characteristic of the unconscious and "primary processes."[2]

Such semiotic mechanisms, however, are not simply the reflection of some nocturnal continent of the unconscious, the poetry of childhood. The world of resonances, connections, and presences operates well outside those spheres. It organizes our imaginary and emotional existence, our most intimate thoughts. It beats in time to an elementary rhythm, provides the fundamental tone for the songs of mankind.

The Semiotic of Territory: Division

Speech in territorial space is detached from living breath and attached to an inert substrate; it is made sedentary through writing. This is true of the first ideograms inscribed on tortoise shells, the tumult of cuneiform on clay tablets. The things to which these signs refer are distant from us in space and time. Signs represent things, make the absent present.

Why is representation the central theme in territorial space alone? Because signs are no longer merely exchanged within a given context but can be separated from their authors, cut off from the living powers to which they clung in the semiotic environment of earth. The changing, living, actual bond that exists among beings, signs, and things is deferred. The separations and borders that divide the territory insinuate themselves into the very heart of the relations of signification: Semiotic division becomes institutionalized.

The state, the hierarchy, and its scribes now come between signs and things. The relationship between sign and thing is no longer one of analogy, proximity, affective continuity, complicated interaction along fluid and numerous channels of meaning. In territorial space, the law establishes names, and words become a matter of convention. Images and sounds no longer function as vectors of force and desire within a living continuum. From now on, the sign represents. It is arbitrary, transcendent.

This transcendence of the sign initiates a situation of absence, a universe flickering between death and life, in which signs and things pursue one another without ever achieving the full presence of being. For the thing is effectively absent, out of our reach. We can never apprehend it except by its name, its concept, its image or percept, that is, through other signs. The thing appears only in the neutralized form, pale and devitalized, of its proxy. It is nothing but an inaccessible "referent." The thing in itself is transcendent. As for the sign, it is still there, but obviously no longer possesses the ontological dignity and immanence of terrestrial objects. It is a lesser being. Imposed by law, transcendent, isolated from the sap that rises from the earth, it is absent in turn. As Lacan has shown, semiological division results in castration.

But signification can never be exhausted by territorializing the sign. Why is it then that nearly all theories of the sign accept the process of semiotic division, whether in its more obvious dualist form or its more subtle tripartite version (signifier, signified, referent, or *vox, conceptus,* and *res*[3])? Isn't it because nearly all such theories are based on the study of texts and the deciphering of dead languages?[4] Because they were born among scribes, clerks, servitors of the state, professors? In such places as churches, schools, and bureaucracies? Because, regardless of the various names by which semiotics has been known since the inception of writing on the banks of the Euphrates five thousand years ago, the transcendence of the sign remains the keystone of political and sacerdotal hierarchies, the secret of our submission to the transcendence of castrated subjects.

The Semiotic of Merchandise: Illusion

Within the commodity space, it is no longer only speech that is cut off from life. Scenes and faces, landscapes and music, rites and spectacles, events of all kinds are indefinitely reproduced and distributed through books, newspapers, photographs, records, film, radio, tapes, and television, cut off from their context of emergence. Multiplied by the media, swept along thousands of pathways and channels, the sign becomes deterritorialized.

Before the appearance of sound recording and the radio, the majority of humanity had heard only the music of its own nation or region, and always within the context of a specific circumstance: work or love songs, seguidillas and bourrées, celebratory songs, religious canticles. Before photography, the cinema, and television, images were attached to specific places, occasions, seasons. The territory separates the object from the sign in order to rearticulate it according to the arbitrary nature of convention, law, and the state. In the commodity space, signs are severed from their origins; they flow without restriction. The division is so effective that the relationship of transcendence no longer serves as a link among the elements of the sign.

Writing provided the means for the analysis of speech, the reification of words, an initial deterritorialization of language. The media brought about a massive and widespread decontextualization of signs that no form of transcendence can regulate any longer.

The semiotics of the territory distinguished the thing from its representation. Within the commodity, or *media* space, things, referents, and originals no longer exist. Money continues to circulate in the absence of a gold standard. A tune heard on the radio or recorded on an LP has never been sung the way I hear it: It is the product of a studio, which only exists within the sphere of the spectacle. The press and television create events, produce a media reality, evolve within their own space instead of sending us the signals of things in themselves. Reference refers only to the mediasphere. The great sign mall, the spectacle,[5] thus becomes a kind of superreality through which every utterance, every image has to pass before it can become effective. The sign's passage through media channels dethrones representation. "As seen on television..."

Within the semiotics of the commodity space, the sign no longer represents; it traces. It no longer functions in terms of a vertically defined division but according to a hundred horizontal lines of circulation.

The sign no longer points toward a meaning or an object; it flows, radiates, diffuses, regenerates, and clones itself, proliferates. It is no longer a representation that has been accredited by transcendence, but a virus attempting to replicate itself, fighting against other viruses to occupy the media space. The circulation of signs is now external to the vital necessity of meaning, as in the hierarchies of transcendence. It swells, enfolds itself, suddenly diminishes in size, establishes a position along temporary heights, chases events, difference, annuls them, seeks new forms of difference, obeys the "laws" of the market, the path of least resistance, in an ever-moving media-driven landscape.

Within the space of reproduction, distribution, and indefinite variation, signs no longer convoke the things they designate, nor the beings that announce them. In the realm of the spectacle, every reality must yield to the sign. Actions, people, works of art. All are signs. And they are processed, reproduced, and distributed as such. Not only does the sign no longer refer to an absent object, it can no longer lead us back to the start of the series, to the "original," since, within the commodity space, the sign is merely a byproduct of the processes of recording, reproduction, and distribution. It is only a sign within the circuit of transmission. Absence triumphs in the midst of abundance. Warhol can silk-screen, Derrida deconstruct, Baudrillard simulate, and Philip K. Dick erect paranoiac universes in which reality is constructed by means of illusion.

The Knowledge Space: Semiotic Productivity

The semiotic of the knowledge space is defined by the return of being, of real and living existence within the sphere of signification. This escape from the world of absence, this resumption of contact with reality should obviously not be understood as a process of objectivization or relation tied to a given signified, a guaranty of signs by means of transcendence. The real is that which implies the practical activity, intellectual and imaginary, of living subjects.

Within the knowledge space, collective intellects reconstruct a plane of immanence of signification in which beings, signs, and things exist in a dynamic relationship of mutual participation, escaping the separation of territorial space as well as the circuits of the spectacle that characterize the commodity space.

At certain rare moments in history or during our individual lives, we reappropriate the creation of signification; we speak for ourselves. The knowledge space could be considered the site of such a continuous resumption of speech, but speech that is effective, capable of changing reality. In the knowledge space active exhalations work together, not to bring about some hypothetical fusion of individual beings, but to collectively inflate the same bubble, thousands of rainbow-tinged bubbles, provisional universes, shared worlds of signification.

The return of the real to the sphere of signification assumes the involvement of living subjects. But it also suggests that the sign space becomes sensible, similar to a physical space (or several of them), which we can enter and navigate, explore, touch, and change, where we can meet others. The knowledge space is nothing more than this virtual reality, this utopia that is already present in patches, stippled, as a potential, everywhere humans dream, think, and act together.

The knowledge space encourages topological metamorphoses. Each intellectual or imagining collective constructs plains, wells, mines, new skies. In each region a different quality of signification is deployed, a different way of signifying.

Within the commodity space sign-projectiles pullulate, searching for targets. In the knowledge space, however, collective intellects organize the mutation of an immense variety of semiotics. The art of the future will no longer forge signs but post-media methods of communication.

Today, by means of hypertext, groupware,[6] interactive multimedia, virtual reality,[7] artificial intelligence and artificial life[8] software, dynamic ideographs,[9] methods of digital simulation, and interactive information systems,[10] collective intellects explore mutant semiotics. In the knowledge space the logocentrism of the territory is no longer legal tender. Coupled to sensorimotor, plastic, interactive, and multidimensional control systems, the image escapes the destiny of fascination traced for it by merchandise and becomes an instrument of watchfulness, knowledge, and invention more powerful than text.

Is this the return to the semiotics of earth? Not entirely, for the significations of the knowledge space have already passed through history, the plurality of languages and worlds, the decoded movement of capital. Within the fourth space, signs no longer refer to some cosmic closure, well-ordered spirals rising from circle to circle, but to wandering and singular lines of significance, to metamorphic spaces of signification.

The collective intellect has already encountered the arbitrary and transcendent nature of the territorial sign. It is thus accustomed to the subtleties of glosses and interpretation. From hermeneutics, a science of territory, it has learned how to dissolve the glue that closely held percept, sign, thing, and meaning together, thus enabling their long sought interaction.

The intelligent community is traversed by the spectacle; it has experienced and benefited from a reality reduced to the sign. It is, therefore, not moved by the nostalgia of authenticity but is firmly committed to the play of artifice, simulation, and unbridled creative imagination.

The collective intellect reappropriates the semiotic productivity that was confiscated by the powers of the territory and the circuits of the spectacle. And even if mankind's only world is found in the element of signification, since it is capable of creating sign systems, the collective intellect can rebuild the world.

Until now we have only reappropriated speech in the service of revolutionary movements, crises, cures, exceptional acts of creation. What would a normal, calm, established appropriation of speech be like, assuming we can so refer to a process that is always in the process of generation? Collective intellects are human environments that encourage subjectivities to continuously differentiate themselves. Construction sites for celibate machines. Armies of spiritual nomads marching along the piers and courtyards of a city of signs in the process of becoming. The surrealists had similar dreams but lacked the technical means to implement them.

Each new way of making meaning creates other subjectivities, other qualities of being. The semiotic productivity of collective intellects is transubstantiated as ontological productivity.

From one space to another, to make real, to give life to something, is to bring things into the daylight of meaning, to make them manifest by means of signs. For mankind, whatever hasn't been sung doesn't exist.

11

Figures of Space and Time

Earth: Immemorial Footprints

Nomads, in their wanderings across the earth, trace a passage: the lines of migration caused by herds of reindeer or bison, the periodic appearance of sources of water, the fertile places that shift with the seasons.

Our gods and ancestors during their wanderings, named the mountains, rivers, rocks, and tall trees. The memorable sites of the nomadic earth are its ossuaries, the remains of gigantomachia, heroic battles. There are signs strewn about of the forbidden love affairs between women and gods. The petrified forms of the animals of myth are outlined against the sky of prehistory.

The earth is mankind's memory, its landscape the map of mankind's epics, the storehouse of his knowledge. All space is alive. Songs and stories relate the earth. The earth remembers the dreamtime, the time of origins, which is always present. The earth will perish with the gods if its songs are not sung again, new voyages not undertaken, its spoor abandoned.

We set off once again across the earth, following the tracks of our ancestors. We return to the same places, singing the song of earth once more. And the past, always with us, lives on. Here, space is crossed by forces, punctuated by high places, intensities, centers, marked out with forbidden regions. The earth engenders a space of qualities, intensities, a memory-space, a narration-space. It is the incarnation of a collective subjectivity in a cosmos.

When innovation arrives, as it always must, the clans ensure its integration with the time of origins. Invention is a form of reminiscence. So that, by a kind of cosmic return, becoming on earth nourishes eternity. Time, on earth, is immemorial.

The immemorial earth carries its time with it; it is always already present, never past. We are on earth when we travel to the Moon. Pilgrims, travelers, adventurers, and poets awaken the earth. Every inhabited space reconstructs the earth.

Territory:
Closure, Inscription, History

"The first person who enclosed a piece of land and proclaimed, 'This is mine,' and found people who were simpleminded enough to believe him, was the true founder of civil society."[1] The founder of civilization, of the territory. Except that now the emphasis is not on appropriation but the very gesture of closure, the work of excavation, digging ditches. Foundation is the very act that creates territory. Every time we build something, in the sense of engineering and architecture, in the sense of building for the long term, we extend the empire of territory.

OVERVIEW OF THE FOUR SPACES
IDENTITY, SEMIOTICS, SPACE, TIME

	Earth	Territory	Commodity space	Knowledge space
Point of irreversibility	70,000 BC	3000 BC	1750	2000?
	Relationship to the cosmos	Relationship to the territory	Relationship to production and exchange	Relationship to knowledge in all its diversity
Identity	"Microcosmos" Affiliation Alliance	"Micropolis" Property Address	"Small house" Skills Employment	"Polycosmos" Distributed and nomadic identity, in comparison to identities of belonging Quantum identity
Semiotics	Presence Mutual participation in signs, things, and beings Correspondences	Absence Division and articulation between sign, thing, and being Representations	Illusion Break between sign, thing, and being Propagation	Semiotic productivity Involvement of beings in worlds of signification Mutations
Figures of space	Footprints Memory-space	Closure Foundation	Networks Circuits Urban	Metamorphic spaces emerging from collective becomings
Figures of time	Immemorial	History "Slow" time, deferred time, engendered by spatial operations of closure and foundation	Real time Abstract and uniform time of clocks	Reappropriation of subjective temporality Adjustment and coordination of rhythms

Foundation[2] is the genesis of a space and also the inauguration of an era, not by inscription on a place that is already present, nor through establishment within a pre-existing span of time, but by the creation and extension of a territorial space-time sustained by the perpetual work of foundation and refoundation. Foundation and empire.

The peasant defines the borders of his field, forages on it, works and plants it. The king digs ditches, builds walls around the city, erects his palace in the center. The priest delimits the sacred space, erects the columns of the peristyle, tends to the holy of holies within the secret heart of the temple where the idol dwells, an altar, or absence. The scribe prepares the clay tablet, the papyrus, the vellum, the page, and inscribes his text within its margins. It is always the same, doubled gesture of territorialization: the delimitation of an area and the construction, planting, or inscription on its consecrated surface.

The enclosed field shelters domesticated animals, after they have been selected and marked. Borders prevent the passage of nomads, cut across their footprints. The state encloses the wanderers. Canals and roads channel movement. Customs, counters, gates, locks constantly reestablish what is in and what is out. Among the scribes, exams and competitions create barriers around knowledge.

Time follows. For the territory creates time from space. Like the Roman limes, the Great Wall of China, all fortifications are ramparts against effacement and oblivion, attempts at duration, endurance. Time begins to flow with the foundation of the city, the initiation of the dynasty. The labored field, sown with seed, echoes the coming harvest through the play of difference, a deferral that creates territorial time. Agriculture brings with it the risks of duration, delay, and storage. Barns, silos, warehouses, cellars, buried treasure, the anticipation of bad harvests, the gamble on the future.

Storage of meaning as well. The page, scratched and strewn with signs, anticipates its subsequent interpretation and commentary. Writing endures, perpetuates discourse. Speech vanishes in thin air; writing remains.

There is a before and after only because there is an inside and outside. The territory secretes the linear time of history. History doesn't travel faster than the immemorial. It is a different speed, a different kind of time, the torpor of territorial space.

Commodity: Circuits, Real-Time

Once deterritorialized, people, things, technology, capital, signs, and skills are renewed, endlessly circulating within commodity channels. Commodity strategies no longer build ramparts; they create networks, organize short circuits. Networks of communication, transport, distribution, and production are inextricably wound together, weaving a space of circulation.

Capitalism is planetary. The course of events on one continent affects the most insignificant activity on another. Circuits and channels of exchange have created global interdependence.

Circulation devours, covers, obscures, buries, and deafens the city. It pierces, tears, and dissects the countryside. The territorial distinction between the city and the country is no longer relevant and gives way to urban life. Along the edges of mines, around ports, up and down railroad tracks, urban life grows and metastasizes. Crisscrossed by roads and highways, hollowed out by subway lines, creased by bus and trolley routes, overflown by aircraft, a knot of roadways, canals, and networks, marked by the detritus left behind by multitudes of intersecting currents; the urban space is a city whose center is everywhere and whose circumference nowhere.

Warehouses, industrial zones, the forgotten housing development lost between a beet field and an airport, suburbs, new towns, commercial centers, all those places that are no place, unconsecrated, without history, agglomerated by new circuits, will never constitute a city. They are peopled, but we are incapable of inhabiting them.

Our mammoth cities are no longer cities. World-cities, inflationary megalopolises, black holes strung across the planet, monstrous attractors: Lagos, Calcutta, Cairo, Mexico City, Los Angeles, the unlivable city, city of automobiles, Tokyo, city of chaos.

But it is not the urban that determines the spatial outlines of commodity, for it is merely the trace of the acceleration of moving objects (people, things, capital, signs, etc.) across territorial space. Abandoned earth. Deterritorialized territory, the urban reveals the flip side of the circuits of communication, networks seen from the outside, when we are disconnected. The commodity space lives entirely within its circuits: on the highway or train, not on the landscape we cross, in the airplane, not in the village near the airport. It is the intensive space of moving objects, a movement-space in which we take pleasure in velocity, acceleration, ubiquity, instantaneous contact. The individual links of the network continue to circulate in other networks: the television in our pocket, the head-phones on our ears, the laptop in its case, the portable fax machine, the mobile phone. Objects in motion within objects in motion combine their velocities, exchange messages, intersect a moving, relativistic space, in which everything moves in relation to everything else, where distance means nothing and speed everything.

Einstein's theory of relativity is obviously the product of the movement-space of merchandise, as witnessed by the intellectual experiments used to illustrate it: clocks, trains, elevators, spaceships, inside one another, their velocities compared. Kanban. The theory of zero inventory cancels out the territorial interplay of future and duration. Deferral vanishes in the just-in-time world of industry as it does in the world of live broadcasts used by the media. In the realm of telecommunications and information technology, real time designates the immediacy of transmission, calculation, and response, the instantaneous processing and presentation of information. On the horizon of acceleration, in the eye of the hurricane of speed, *real time*, motionless, drives the space-time of commodity. Real time is the reality of commodity time, its entelechy, its ideal: a time that is no longer sequential but parallel, no longer linear but instantaneous, the time of simultaneity, the limit of acceleration.

The most advanced form of real time occurs within organizations. From flexible manufacturing systems to groupware, digital networks have brought about the dematerialization of organizational structure. The ultimate deterritorialization: organization charts, production procedures, and administrative architectures are transferred to software and thereby mobilized and tamed. The virtual enterprise adapts in real time to the transformation of the market. In this sense it is similar to the knowledge space. But we can't get there simply by going faster. We need some sort of qualitative leap. Different velocities, different intensities will animate collective intellects.

Knowledge: Subjective Time, Interior Space

The time we are accustomed to seeing on our watches and calendars and schedules, mechanical time, the time used to measure velocity, synchronize actions, and coordinate organizations, all these notions of time are aligned with the regular movement of the stars.[3] Transcendent time falls from the sky to subjugate the earth. It is the unifying time of the territory. The clock, the giant clock in Chaplin's "Modern Times,"[4] is synchronized to the rhythm of infernal machines, the time of wage labor, transport time, down time, lost time, the time of boredom. The territory has homogenized, domesticated at great cost, living and subjective time. Commodity expropriates subjective temporalities.

Within the knowledge space collective intellects secrete their own periodicities, individuals reappropriate their subjective temporalities.[5] The knowledge space abolishes deferral, but not in accordance with the modes of commodity space, by accelerating to real time, because real time is still indexed to the clock, to an external time. The knowledge space annuls deferral by changing the reference system: It is nourished on internal time. Rapid, intense durations, entirely contained in vitality, the calmness, the tranquillity of collective maturity. Such velocity, such slowness, has no relationship to the clock or calendar. It refers only to itself. It is a quality of being. In the knowledge space, time flows from a variety of living sources that blend together. Different times bubble up and call to one another like musical rhythms.

The deferred contradicts internal rhythms. Bureaucratic delays break our momentum. The mad velocity of real time suppresses and destroys anything that attempts to grow slowly. Still, the collective intellect does not ask for everything, not at once. The fourth space rejects the blind rule exhorting us to "life without boredom, pleasure without fear."*

In response to the criticism that this is a utopian vision, I would say that, yes, the knowledge space is utopian, but it is an achievable utopia. There is always the danger that these subjective temporalities will close in

---

*The original is "Vivre sans temps mort et jouir sans entraves," literally "live without dead time, come without impediments," a well-known Situationist slogan that could be read on the walls of Paris in May 1968.

upon themselves, that they will end up in isolation, autism, idiocy (in the etymological sense of the word "idiot," as someone special or peculiar). To counteract this I would offer two different strategies, one defensive, the other offensive.

With respect to the defensive strategy, we need to remember that collective intellects do not abolish previous spaces. They are content to deploy other qualities of being, other temporalities. Within their spheres, territorial and commodity relations will continue to operate. With respect to the offensive strategy, I would emphasize that the knowledge space offers the technical possibility of constructing personal temporalities for the creation of a collective subjectivity and ensuring that collective, emerging time affects individual subjectivities. By following their own rhythm individuals will not be condemned to isolation.

In the concept of the knowledge tree,[6] for example, it is the *curricula* of individual apprenticeship that structure the knowledge tree of the community. Yet these curricula are not synchronized with calendar time; the dates on which we obtain our diplomas are not shown. We only take into consideration the subjective element of apprenticeship, its immanent order. Thus, the holes, the spaces, those moments that are empty or hollow only in terms of a transcendent temporality, are not recorded in the knowledge space. An individual's cognitive *blazons*, projections of their curricula on the tree, no longer record a sequence of personal achievement but individual abilities in an emerging collective order, that of the community tree, and thus enable individuals to orient themselves in terms of a specific situation, a context shared by all the other members of the collective intellect. Subjective temporalities compose a common space.

We have seen that the real time of the commodity space assumes the prior domestication of time by the territory, the transcendence of celestial time. In similar fashion the knowledge space would be unable to obtain its autonomy, its irreversibility, without using real-time technologies produced within the commodity space. But these technologies are focused on the interior, used for other purposes, not the construction of simultaneity within an external time, but the adjustment of living rhythms, dynamically related communities of meaning, asynchronous situations.

Space follows. Based on the temporalities of collective thought, spaces of signification take shape and transform themselves, strung with subjective proximities, interior distances.

The Emerging Knowledge Space

Collective intellects appropriate a subjective time because their chronology does not refer to any external, preexisting space, to any physical motion. Their time pushes, grows, becomes. In terms of the opposition introduced by Norbert Wiener in the beginning of *Cybernetics*,[7] this would be a Bergsonian rather than a Newtonian time.

The collective intellect reverses the relationship between time and space that had been established by the territory. In territorial space, time flows from a foundation, which is an operation on space. Duration is based on the presence of an inside and an outside. To control or orient becoming, territorial space makes use of spatial means: walls, canals, gates, drawbridges, bureaucratic labyrinths, never-ending concentric circles of exclusion and belonging.

The collective intellect, however, acts against the grain of territory since it transforms time into space. It is organized around mechanisms that reflect a multitude of events or collective becomings in a dynamic and qualitatively differentiated space (obviously this is a sign space, a cartographic not a physical space).

The territory attempts to maintain borders, hierarchies, and structures. The knowledge space on the other hand is always in an emergent state. It issues from the individual acts and histories that animate the collective intellect. It is never structured *a priori*, but expresses, maps, makes visible the strands of subjective and necessarily unforeseeable durations.

We can create borders from without, divide them up, carve out empires, define sacred precincts, assert rights of ownership, establish customs duties. Were we to do so, however, this would no longer be the knowledge space but the traditional territorial space.[8] We must therefore construct above it a new, unbroken space, free of barriers, continuous, receptive to the multitude of moving figures that trace our collective becoming. The communities of intelligence flee the territory, escape the networks of commodity for a knowledge space that they produce by thinking, dreaming, and wandering.

It is possible that in the face of commodity networks, urban spaces will remain uninhabitable, though in all likelihood they will continue to grow. But above them the electronic conference is taking shape, the infinite discourse of collective intellects. Beyond Los Angeles lies a city of angels, megalopolis of signs, now visible city of the mind, in the night, on the multiple heavens of cathode ray screens.

12

Navigational Instruments

Earth: Narratives, Portolans, Algorithms

The bard sung the voyages of Ulysses. Sirens, magicians, cyclops, sea monsters, strange people, mysterious isles, ill-fated winds, currents: Strange and singular events punctuate the life of the Mediterranean. The hero's encounters organize a space of possible navigations. Ulysses, in the end, returns to Ithaca.

The *Odyssey* is one of the first portolans. From Tyre to Marseilles, Naples to Carthage, the portolan provides all the information needed for the sailor to arrive safely at port. It keeps track of reefs, bottom depth, directions, and sailing time for each segment of a voyage. The portolan can't be used for exact readings, but it helps the sailor identify individual entities such as buoys, seamarks, and lighthouses. Primitive navigation, based on dead reckoning, exchanges the track, or spoor, for expected, though not always successful, encounters that occur within a qualitative space. Ulysses wandered over the seas; his ocean was part of the earth.

The compass is a terrestrial device. Its reference point is magnetic north, a unique point on earth. But the magnetic field fluctuates. The compass constructs a tactile space of proximities: so many days to a given position, so many days to another. Unlike the territory, which constructs diagrammatic, topographic, visual spaces, the earth is known through touch; it reglues or resews, through successive contacts, particular grains, remarkable textures. Homer was blind.

The portolan is an odyssey in which we are the hero, a databank for a thousand-and-one possible stories. On earth, the principal instrument of knowledge is the narrative. Not just the myth or epic, which summarizes primitive genealogies, the identifying figures and beliefs of a people, but the story as a form for organizing practical knowledge. For the portolan contains charts and bits of story, and the navigators will, in turn, become narrators, the authors of different stories. The bard or griot narrates the adventures of heroes who recognize the earth, people it with figures and named places, organize it for those who will come after him. To perpetuate a tale is to transmit an ordered inventory of the qualities and actions possible within a given space.

The abstract structure of the story is the algorithm: the description of a series of actions that occur in a certain order from a starting point to an end point.[1] On earth, technical, especially mathematical, skills are remembered and implemented by means of algorithms. The algorithm does not need to be proved. Like a blank story, it encourages us to become heroes in turn. And because it is efficient, it brings us back to Ithaca.

Territory: Projection Systems

Somewhere between the fifteenth and sixteenth centuries, when Portuguese sailors began sailing around Africa and Columbus was making plans to reach the Indies by a western route, the question was no longer of returning to Ithaca but of adventuring across uncharted seas, discovering unknown lands. Stories and portolans ceased to be of help. Navigation by the stars came into widespread use. Through the use of the quadrant and astrolabe, the captain could precisely measure the height of the pole star or the sun above the horizon. Based on the time, the day, and astronomical tables, he could determine latitude. This is referred to as taking one's bearings. This bearing is no longer an individual point, a landmark, a sign, the color of the water, a whirlpool, a mark on the earth-sea, but an abstract point, the projection on the surface of the earth of celestial coordinates. A position in territorial space.

Most likely it was after the discovery of a space consisting of abstract points that Zeno of Elea was able to formulate his famous paradox on the impossibility of movement. Zeno was right. In territorial space Achilles never catches up with the tortoise. We make no progress within territorial space; we can only determine our position. But on earth, site of numerous wanderings, travels, and voyages, Achilles does catch up with the tortoise. Achilles, like the sailor, lives in both spaces. He takes his bearings and orients himself within territorial space but makes progress on earth.[2] Kafka's surveyor never arrives at the heart of the castle—the center of his novelistic territory—because he never succeeds in making contact with the earth or rising above the territory. The surveyor, as his profession amply demonstrates, is the man enclosed within the territory, an impotent Achilles. Kafka's entire œuvre could be considered a kind of meditation on territorial space.

Because of the new methods of navigation, the sailor can now find his position, at least his latitude, on a sea that no one before him had crossed, on a route that no story had yet marked out. From now on, maps are no longer simply accumulations and extracts of dead reckonings based on distances, angles, orientations, relations of proximity upon the earth. Each point of the map would now be indexed to the sky, pinned in place by the sphere of fixed stars. Soon the oblique and colored rays of the compass rose would disappear from nautical maps, to be replaced by the uniform grid of latitude and longitude. Equator, tropics, poles, parallels, meridians were the first coordinates of the celestial sphere. Although scientific geography was known in antiquity, as shown in the work of Ptolemy, it wasn't until the sixteenth century that the systematic projection of the coordinates of the celestial sphere on the earth became customary. The period of "great discoveries" was the culmination of the formation of territorial space. The earth's surface was no longer organized around the unique figures that our ancestors, heroes, or gods encountered or disseminated during their voyages. In the new space the earth is divided up, enclosed in a net that falls from the heavens. Each point is assigned a set of coordinates, an address, even though it remains unnamed. The story or algorithm is replaced by a system (of which scientific geography is only a particular manifestation). The former organized lines of adventure, series of operations stabilized through experience. For the system, all space lies open, all operations are possible. The system enables us to find our way where no one had gone before, to successfully complete an operation undertaken for the first time, since all of space has now been recorded.

What is a territory in the order of knowledge? The projection of the heavens upon the earth. The imposition of a transcendent order, celestial, totalizing, on an immanent and incomplete space of singularities. The nomadic earth immobilized by the sky.

Commodity: Statistics and Probabilities

In the commodity space there is no longer any possibility of fixing positions—beings, signs, or things—within a system. Everything is in circulation; everything in a state of continuous flux. It is no longer simply his passage across an immobile territory that the merchant must account for, but gross changes to a large number of data, a heterogeneous and diverse flow of events.

At each instant, at each station of the commodity network, throughput, tonnage, prices, sales figures, exchange rates, media consumption, deaths and births can be counted, measured, and recorded. But there is no fixed sphere looking down on the raging sea of numbers. The only resource left is the uncertain sky of economic conditions. Can decisions be made on the basis of such disordered agitation? How can we represent the current status of a situation, discern the general outline of events on the basis of such movements?

Statistics and probability didn't make their appearance until the seventeenth century, and their development accompanied capitalist expansion. They were used to draw up the moving maps of the commodity space. Indexes, rates, percentages, averages, trends, curves, ratings, histograms, and pie charts aggregate a heterogeneous crowd of information, summarize a multitude of individual histories, supply the global profile of a dispersed plurality of events.

But as indispensable as these instruments[3] are, in the absence of other means of orientation in the commodity space, they quickly reveal their drawbacks. Individualities are no longer fixed by a system, as in territorial space, but lumped together in a single mass. The statistical profile masks individual features. Processes are equated with molar equilibria, with states. Etymologically, statistics is, strictly speaking, the science of states. Dynamic configurations[4] are reduced to averages; strategies are subjected to the laws of large numbers, to probabilities. Qualities are reduced to quantities.

Knowledge: Cinemaps

The instruments for navigating territorial space are designed to work with durable spaces and objects. The inertia of the territory is strong, its beacons fixed. Geometry has idealized this slowness of movement in an eternal and abstract space. Economic space is moving, relativistic. Here the terrain itself moves and is transformed. Commodities circulate, their prices fluctuate, have different values from one place to another at the same instant, systems of production evolve, etc. The astronomical sky is succeeded by the meteorological sky. Yet knowledge is born, runs, and dies at an even faster pace than commodities. The value and organization of knowledge is closely linked to a changing cultural, social, and professional context. It depends on the specific goals and objectives of the individual at a given moment. Only a contextually vital interpretation gives meaning and value to knowledge.

The main problem in the knowledge space is to organize the organizing, objectivize the subjectivizing. Knowledge about knowledge is based on an essential circularity, one that is primordial, ineluctable. Our understanding of understanding is *ipso facto* a transformation of knowledge, a perpetual deviation, a dynamic reorientation that is continuously reactivated and reevaluated.

The cartography of the knowledge space cannot be based on statistics, which are purely quantitative. Together with Michel Authier, I have suggested a new type of instrument of orientation and navigation, especially designed for the knowledge space, which we refer to as a cinemap. A collective intellect navigates within a moving informational universe: A cinemap is the product of this interaction. On the cinemap the informational universe (or databank) is not structured *a priori*, in keeping with some form of transcendent organization similar to that found in territorial space. It is not standardized by the use of statistical averages or distributions, as in commodity space. The cinemap integrates the qualitatively differentiated space containing the attributes of all the objects in the informational universe. The topological organization of this space expresses the variety of relations experienced by the objects or actors in this universe. Each object or group of objects in the universe under consideration can be visualized on the cinemap. We can measure the distances between these objects, or between a given object and an attribute it doesn't yet possess, for example.

On this interactive cinemap, every quality, every singularity can be situated and visualized. Points on the cinemap are not abstract units, identified only through their coordinates, as on the maps used in territorial space. Each point on the cinemap is a different attribute, a particular quality, shown by an icon, a unique sign. The cinemap is a moving mosaic in a state of permanent recomposition, in which each fragment is already a complete figure but one that, at each instant, only assumes its full meaning and value within the general configuration. Behind each sign-point, hypertexts and messages provide additional information, encourage deeper investigation, detail the resources needed for navigating the knowledge space. The cinemap enables us to explore a dynamic macro-singularity consisting of many, individual singularities.

The objects or actors in the informational universe continuously transform themselves, lose and gain attributes. Thus they move across the cinemap. But since the cinemap expresses the relations among informational objects, it evolves and restructures itself at the same time as the actors modify themselves and new actors and new attributes appear. The continuous film representing the transformation of the cinemap is an important source of information, which transcends its instantaneous state, its status at a given point in time.

The specific use a community makes of the cinemap and its message system is useful in evaluating the objects and attributes of the informational universe, according to a range of criteria. Values, which emerge from collective use, are made visible on the cinemap by means of color, which in turn helps promote further use and investigation by the collective intellect. Each member of the collective intellect can find his individual location on the cinemap. In this case, two possibilities present themselves.

In the first case the informational universe models the collective intellect itself, the diversity of its attributes, its behavior. Here the individual is one of the object-actors who help structure the cinemap. Obviously the individual can find his position on the map. He is indexed by a constellation of icons. His signature appears on the cinemap of the collective.

In the second, the object-actors of the informational universe do not model the members of the collective intellect but other data. Here, individuals are still recorded on the cinemap according to their preferences, their interests, their relationship to the objects of the informational universe. Their position on the cinemap, indirectly assigned this time, can again be visualized by a distribution of attributes, a configuration of sign-points. Based on their position, they can continue to develop their strategies of navigation.

As with the maps used in territorial or commodity space, cinemaps have an immediate practical consequence: finding our position, preparing itineraries and strategies, coordinating with others through the use of shared representations, etc. The cinemap enables us to read a situation, a dynamic configuration, a qualitative space of signification shared by the members of the collective intellect or collective imagining. By using a cinemap a group can shape itself into a collective intellect. Reciprocally, the cinemap is a virtual reality, a cyberspace created by the exploratory activities of a collective intellect within an informational universe. Cinemaps will be part of the navigations of the collective intellect in its knowledge worlds; they will relate its discoveries. Cinemaps will continuously map the knowledge space, express a collective subjectivity. By providing it with the proper instruments for navigation and orientation, they will help shape the knowledge space, make it irreversible and autonomous.

2001: the Odyssey of Knowledge.

13

Objects of Knowledge

Within each anthropological space, the most important objects of knowledge are the figures specific to that space. These objects are constituted by the instruments of navigation for the space in question. Anthropological spaces are capable of self-reflection using the tools they have made.

The Object of Earth:
The Eternal Becoming-Beginning

On earth the object of the story is an origin. The clan understands a thing, an institution, through its birth, when it is encountered for the first time. But stories are transmitted, perpetuated, related in cycles, varied. The idea of the story is already the result of something made permanent through writing. On earth there may not be any stories, only an indefinite chain of narrations undergoing a continuous process of drift and correction. The true object of narration is not an origin, the fixed point of departure, as in the linear histories of the territory, but the immemorial. The object of terrestrial knowledge is an eternal becoming-beginning.

The algorithm, without place, undated, anonymous, represents such a timeless becoming-beginning, already fulfilled, always awaiting fulfillment—the algorithm, or ritual.

The Object of Territory: Geometry

The object of the territory is the measured earth, divided into sectors: geo-metry. According to legend, Thales, emblematic figure representing the birth of geometry, discovered the theorem that bears his name by attempting to measure the height of the pyramid of Cheops from the length of its shadow. Thales demonstrated that the ratio between the height of the pyramid and the length of this shadow must be identical to the ratio between the length, which was known, of a stick stuck in the ground and the length of its shadow (the height of the sun being identical in both cases). Michel Serres has often commented on the significance of this "primitive scene" from geometry.[1]

If the territory is a measured earth, one that has been *geometrized*, the birth of geometry reveals one of the genealogies of the territory. Thales's story illustrates our definition of the territory as an earth immobilized by the sky. First, the earth is measured by the astronomical sky, the sun and its inclination, from which the shadow results. Second, the nomadic earth of the desert is measured by the sky of the state, represented here by the tomb of the despot. The summit of the state is always celestial or divine: the emperor of China is the son of heaven, the pharaoh and Louis XIV are sun-kings. The pyramid is itself an image of the bureaucratic and sacerdotal state. And finally, the earth is measured and immobilized by the sky of ideas. Thales's theorem is one of the first geometric hypotheses to be proven and inaugurates the reign of mathematical idealization, which, in Plato's model, was made up of intelligible and immutable forms.

The exact sciences are based on geometry, primarily classical physics, which "territorialized" the cosmos. "The book of nature is written in geometric characters." Writing and geometry. In one sentence Galileo evoked the instruments with which the territory is formed.

The Commodity Object:
Flux, Fire, Crowds

The characteristic object of the third space is obviously the production and consumption of commodities. Economy is in some sense the geometry or cartography of the space opened up by generalized commodity exchange. But before it can reflect on itself, the third space must do more than simply describe the variations and movements of money, goods, and people. It must also be capable of apprehending the accelerated, chaotic, uncertain world exposed by deterritorialization. The objects of the commodity space are not only those of economy but also everything that is distributed, flows, circulates, is transformed and perishes, everything that feeds its machinery and streams through its channels.

It was in response to this sense of movement and exchange that thermodynamics, the science of the transformation of energy, was born in the nineteenth century, amidst the fire and steam of the industrial revolution.[2] Along with the development of telecommunications, radio communications, and electronics, the first part of the twentieth century developed an information theory that was conceived on the basis of the circulation of signs within networks. The process of encoding, decoding, and translating information, and the struggle to reduce the noise that weakens and destroys messages in the communications channel were the two major problems facing the new science.

For both the mathematical theory of information and cybernetics, the quantity of information carried by a message is inversely proportional to the probability of appearance of that message.[3] According to this definition, the quantity of information will be maximized if the message is completely random—which obviously clashes with common sense—and the quantity of information will approach zero if the message is highly redundant, if it contributes little to reducing our uncertainty about the state of the reference universe. The information that is "sensed" is thus situated, the majority of the time, "between crystal and smoke,"[4] that is, between redundant order and chaos. Information is apprehended by means of the mathematical instruments created by thermodynamics, ultimately, statistics and probability.

Much has been written about the analogy between noise in information theory, which corrodes, disorganizes, and punctures messages, and entropy in thermodynamics, which blurs distinctions and minimizes tension. Since noise not only destroys messages but also creates new information, the transition between energy and information provides the key to reversing organizational disorder, one of the mechanisms of self-organization. These speculations followed the economic changes that have been taking place since the end of the Second World War: Communications networks control the distribution of energy, the management of signs controls the production of material goods.

The theory of information, however, is merely a mathematical interpretation of the transmission and circulation of messages. Admittedly, it has been refined considerably and has resulted in some fairly subtle transpositions. Nevertheless, using the quantitative instruments and concepts it has developed,[5] no one has ever succeeded, even remotely, in analyzing signification, in the most ordinary sense of the term.

As the anthropological space of networks and deterritorialized acceleration continues to grow, physics becomes increasingly preoccupied with processes rather than laws, disequilibrium rather than states.[6] The deterministic chaos[7] and fractal objects[8] studied by the natural sciences echo the fads, erratic behavior, and randomness that now characterize the human world.

Frequencies, harmonics, vibrations, the roar of motors, messages, loudspeakers, the clamor of large numbers, sales, background noise, chaos, media hubbub, the Big Bang, big metal ears pointed toward the edges of the universe—just as the earth is tactile and the territory visual, the third space, like an echo chamber, resonates with the vibrations of all sounds.

The Object of the Knowledge Space: Signification and Freedom

Collective intellects and their worlds are the characteristic objects of understanding in the knowledge space. Collective intellects are human communities in the act of self-communication and self-reflection, involved in the permanent negotiation of relations and contexts of shared signification. Their worlds consist of their resources, environments, and cosmopolitan connections with beings, signs, and things, their involvement in the various cosmic, technological, and social machinery that crosses their path. There is nothing stable or objective about the world of a collective intellect. It is the result of discoveries, developments, fluctuating uses and evaluations, repeatedly reiterated. This world drifts and transforms itself in synch with the metamorphoses of the collective intellect.

The characteristic object of the knowledge space is not mankind in general, nor the object of the humanities or social sciences, but a figure specific to the knowledge space, which can only be apprehended with the conceptual and technical instruments of this space. In this sense the most appropriate candidate for the object of understanding of this fourth space would appear to be cognition. In fact, since the Second World War, the growth of the cognitive sciences has accompanied the rise to power of the knowledge space. The sciences and technologies of cognition have developed effective computational tools and conceptual instruments for use in the fourth space. But although it has indeed contributed to a genuine anthropological mutation, cognitive science has not yet fulfilled its true potential. It has been limited to human intelligence in general, independent of time, place, or culture, while intelligence has always been artificial, outfitted with signs and technologies, in the process of becoming, collective.

In *The Technologies of Intelligence*, I suggested that the scope of the cognitive sciences be expanded and outlined a program of "cognitive ecology." The system for the production and distribution of knowledge doesn't depend on the individual features of the human cognitive system alone, but also on collective methods of organization and the instruments with which information is communicated and processed. Thus, cognitive ecology would focus on studying the interactions between the biological, social, and technical determinants of understanding. But it would remain a part of the social sciences or humanities, whereas I am interested in something completely outside the realm of the humanities.

Were they to take into account the specificity of their object, the humanities' primary concern would be freedom and signification. The concept of signification should here be understood in its broadest sense. It refers less to the way in which a sign refers to its object than to the relationship between an event and a context. The signification of an entity that falls within the human domain can only be understood in terms of its relation with a dynamic configuration and can't be reduced to reference alone. We might say that human beings, as flesh and blood men and women rather than physical objects or living things, evolve within spaces of signification, spaces that they do not simply occupy or cross but which they also produce and transform.

Along the same lines, the becoming that should be the subject of the humanities is obviously not a simple succession of events in time, but the creation and destruction of new forms and new worlds of signification, which initiate their own temporality. The event is not simply dated, it is potentially dating as well, that is, capable of affecting and modifying the configurations of meaning inhabited by mankind. The concepts of "occasion" and "opportune moment" cannot be dissociated from the games and strategies we are so often involved in. They are part of a contrived time, one that is moving, scattered with singularities, one that fashions the fabric of our lives.

Evaluation. Evaluation provides the human landscape with its sense of variation and difference. It is precisely when we ignore the good, the beautiful, the useful, the precious, etc., in favor of some feigned objectivity, that objectivity fails most abjectly. Until now economy has been the only science that has succeeded in analyzing value, even though questions of its origin or source continue to divide economists. There are clearly other methods of enhancing value aside from those characteristic of commodity economics, yet we are still incapable of employing them in any useful way.

The humanities must also be capable of taking singularity into account. The objects which they analyze are of less interest as passive entities subject to general laws than as agents that participate in the creation, maintenance, and destruction of space, temporality, and value in the world of signification. In terms of meaning and freedom, there is always a relation of co-definition between a singular entity and a collective or community. The singular entity can never be thought of separately, but always in terms of its relation, not with some preexisting norm or universal law, but with other singularities, with which it blends to produce emerging and transitional laws. The well-known problems concerning the relationship between the one and the many, or the universal and the particular, are here expressed as the question of coordination or collaboration between autonomous singularities.

We can now ask to what extent do the instruments characteristic of the humanities (structures, statistics, computational models) enable them to apprehend, in some functional manner, the concepts of signification, event, value, and singularity that have just been evoked?

Structure. Inspired by mathematics, methods of structural analysis made their appearance in linguistics at the beginning of the twentieth century. They were soon applied to other fields, primarily anthropology. By considering the internal relationships among systems of discrete elements (opposition, substitution, etc.), structural analysis revealed an important dimension of signifying phenomena. However, the analysis of structure tells us nothing about its genesis or the way in which its elements are instantiated and assume meaning within a given context. Structural analysis only provides insight into static aspects of the paradigm. While structural semantics certainly exists, the concepts of situation, event, occasion, and action that define pragmatics (and signification in the strongest sense of the word), are beyond its reach. Moreover, such structures are generally constructed manually, following some lengthy, and often questionable, effort of interpretation. Rather than depend on the insight of specialized researchers, it would be preferable if such structures themselves were to express the process of interpretation of the subjects they are concerned with (speakers in linguistics, cultures in anthropology, etc.). Structure is a modality of the system. Structuralism is therefore an instrument of knowledge characteristic of territorial space, manipulated by professional interpreters, who distribute its objects in accordance with transcendent coordinates or categories, and reconstruct, in theory, all possible cases. While we can postulate a grammar of the human world, its systems of affiliation, its technologies, its myths, etc., it is clear that such a grammar will remain inert as long as it is isolated from the pragmatics that engender, animate, and give it meaning.

Statistics and probability.   As stated earlier, statistics and probability first appeared during the seventeenth century in connection with the development of the monetary economy in general and insurance in particular. They continued to develop with the growth of the management role of the government at the beginning of the nineteenth century. In spite of their obvious utility, such tools have always described the objects under analysis in terms of large numbers. Probabilities, averages, percentages, standard deviations, and variance, all tend to merge individual features, subjecting them to the standards they helped develop. They provide us with a poor understanding of the nature of singularity as such, of events and their effects, dynamic configurations, anything, in fact, that is part of the world of signification in general. Although very common in sociology, demography, game theory, and other branches of the humanities, statistics and probability are instruments of the commodity space, and incapable of analyzing freedom and signification.

Computation.   A wide range of scientific tools has been developed since the thirties, whose principal characteristic is to provide an operational and formal translation of concepts that have been ignored by classical physics but are necessary to an understanding of life and thought: information, finality, calculation, reasoning. This approach is reflected in the rich tradition that includes Turing machines, Shannon's information theory, the theory of automata (McCulloch, von Neumann), cybernetics (Wiener), etc. The formal instruments in question have had wide application in engineering (information technology) and biology. Aside from the wealth of metaphors they have supplied to anthropologists, sociologists, family therapists, and management specialists, the primary uses of these formal tools in the humanities have been in the cognitive sciences, namely linguistics (especially Chomsky's generative grammar), cognitive psychology, neuroscience, and artificial intelligence (to the extent that the latter is not understood as a form of engineering but in the wider sense of attempting to understand the mechanisms of learning, memory, reasoning, and perception).

Are the instruments supplied by information theory and automata capable of accounting for the historicity, signification, and singularity that constitute the human world? Information, as understood by Shannon's theory, has very little in common with our conventional interpretation of the word. It does not take into account the sense of messages but only their probability of appearance, which brings us back to our earlier discussion of statistics and probability. Computational linguistics does an excellent job of handling syntax but tells us little about the meaning of words in context, perhaps the most significant aspect of communication. Computational approaches to intelligence construct partial models out of small modules of cognitive functionality, but what claims can be made for an intelligence that doesn't understand the meaning of its actions?

While some may feel that I am being unduly harsh on formal instruments and disciplines that have proved their usefulness, the failure to point out their limitations would delay the creation of a knowledge space in which collective intellects are capable of self-recognition. The object of knowledge of the fourth space is beyond the reach of the humanities because its subject—the collective intellect—does not claim to produce objective knowledge of itself or its world. Neither signification nor freedom are objects that can be externally apprehended. Dynamic configurations, initiating temporalities, evaluation, negotiations among singularities within the human sphere, can only be apprehended within the spaces in which they unfold. They must first affect us before we can understand them. Here, knowledge is inseparable from subjective implication, the formation of the subjects of knowledge by their objects. In the fourth space, the collective subject of knowledge is immersed in its object, that is to say, in its world, the living environment on which it depends and which it helps construct.

In the knowledge space, each discovery is a creation. The thinking community that wanders across the knowledge space cannot be located by any form of celestial mapping, or dissected by any system of transcendent categories. As a center for producing and evaluating qualities, it cannot be reduced to the fluctuation or distribution of quantities. Inhabited by individual signatures that rise, blend with one another, and then disappear, the space of sapience is olfactory. The knowledge space is a repository of flavor.

In the knowledge space, knowledge means redefining an identity, locating and modifying dynamic configurations, yielding to a dialectic of continuous evaluation, decision, and reevaluation of the criteria of evaluation.

The instrument of knowledge in the knowledge space—the cinemap or virtual world of shared signification—does not objectivize. It helps support a perpetual process of creation of signification. A tool of self-knowledge and enhancement of potential, it encourages us to exercise our freedom.

Cinemaps enable collectives to singularize themselves, dynamically identify themselves in terms of the worlds they help bring into existence. For example, by using knowledge trees,[9] collective intellects can continuously update their classification of knowledge while they manage their skills and learning abilities in real time. Using health cinemaps, collective intellects can increase their medical skills, manage pharmaceutical and medical resources, combine epidemiology with the prevention, training, and empowerment of individuals. Using language cinemaps, collective intellects can prepare semantic diagrams of their communications, enabling readers to select the texts that interest them, providing authors with a writing tool, and enabling everyone to locate messages within an evolving global context.[10]

Knowledge trees generate dynamic epistemologies as a byproduct. They are primarily a form of self-management for learning and training. The health space supplies epidemiological data only within the context of communal self-medication. The language cinemap serves primarily to locate itself within a communications network, a documentary corpus. It so happens that it is also the instrument of a text pragmatics. A dynamic cartography of personal relations in a collective intellect would produce an auto-sociology only as a byproduct of its activity, etc. The intelligent collective doesn't analyze itself to understand itself: It understands itself because it lives and only understands itself by living. Within the knowledge space knowledge no longer objectivizes but subjectivizes, on the basis of a subjectivity that is plural, open, and nomadic. In the knowledge space the preferred object of knowledge reflects the eternal becoming-beginning of the earth, the perpetual resumption of becoming of the collective intellect and its world.

14

Epistemologies

Earth: Flesh

On earth, the subject of knowledge is the clan and its members, the clan that apprehends and transmits, from one generation to the next, thus sustaining the duration of knowledge. Such knowledge is distributed throughout the collective subject. It is immanent in its being, its life, its practices. As a result, and assuming that the division of labor and function is not pushed too far, a small group, or even a single individual, can master the sum total of knowledge.

The substrate of knowledge, the encyclopedia of the earth, is the earth itself. But it is our physical bodies, experienced and capable, our memory and repeated actions that bear the world's knowledge. On earth, when an old man dies, a library goes up in flames.

Terrestrial knowledge has been incarnated. Intuition discovers and the flesh remembers. Phenomenology and

OVERVIEW OF THE FOUR SPACES
THE RELATIONSHIP TO KNOWLEDGE

	Earth	Territory	Commodity space	Knowledge space
Navigational instruments	Narratives Algorithms Portolans	Projection of the sky on the earth Systems Maps	Statistics Probability	Virtual worlds Cinemaps
Objects	Becoming-beginning Ritual	Geo-metry "Laws" of nature Stability	Flux Fire Crowds Objects of the "humanities"	Signification Freedom Dynamic configuration of collective subjects-objects-languages Renewal of becoming of the collective interest
Subjects	The elderly	Commentators	Scholars	Intelligent collectives Humanity
Substrates	The body of the community	The book	From the library to hypertext	The cosmopedia
Epistemologies	Empiricism Phenomenology	Rationalism Transcendental idealism Scientific method Paradigms	Theory of action and networks (operativity, technoscience) Theory of narrative (modeling, simulation, scenario) Theory of art (artificial intelligence, artificial life)	Social practice of knowledge as a living continuum in constant metamorphosis Construction of being through knowledge Philosophy of implication

radical empiricism may be the theories of knowledge that best correspond to the first anthropological space.

Territory: The Book

In territorial space the subject of knowledge is the caste of literary specialists, hermeneuts, the guardians of systems. Territorial knowledge is a reserved domain, confiscated and transcendent. It is shut off from the outside world like a sealed book. Closed, cut up, divided into an elaborate hierarchy of concentric circles, successive levels that are increasingly difficult to access. Knowledge is the image of the territory. Surrounded by walls, it leaves the peasants, the ignorant, outside. Once inside the gates of the city, however, the precincts of the temple, the secret chambers of the pyramid still remain.

The book contains territorial knowledge. Not books, not the library, but the Book: the Bible, the Koran, sacred texts, the classics, Confucius, Aristotle ... The territory reads and writes only to interpret the book, the infinitely interpretable book or utterance that contains all, explains all, can interpret all. The book or the system. Hermeneutics or deduction: expanding empires, histories that roll through time, foundation, figures of territorial space. In turn the system is an architecture, well-founded, like a pyramid, a fortress.

Epistemology, like semiology, appears to be primarily focused on territorial space. The dialectics of theory and experiment are clearly part of a territorial diagram: The theoretical sky overlooks and organizes forms of practice and experience that are nomadic, terrestrial. When we say that practice or experience have an effect on theory, strengthen or weaken it, we remain within the same vertical schema. The terrestrial shadows of sundials and gnomons[1] have been used for centuries to under-

stand and measure the heavens. But it is still the sky that
measures, projects, and contemplates itself. The earth is
present merely to exalt the heavens, its laws, its transcendence, its universality. Epistemology diffuses the music of
the spheres rather than the song of the earth. It is true that
territorial knowledge is supported by a dialectic between
heaven and earth, for without experience there is neither
practice nor theory. But it is always theory that determines truth.

The opposition between rationalism and ordinary
empiricism, like that between the transcendental subject
and phenomena, is part of the same territorial space, the
same vertical map. The history of science is conceived as a
succession of paradigms,[2] it defines the surface or hori-
zontality of the territory: foundation, stabilization of a
field or a domain, decline, reversal by a new, more powerful paradigm. It is the history of empire, or simply History. Paradigm-based epistemology merely follows the
thread of territorial time.

Commodity Space: Hypertext

In the commodity space the subject of knowledge is
the military-industrial-media-university complex, generally referred to as technoscience. Far from remaining the
guardian of a restricted temple, technoscience is an engine that pulls along with it the accelerated, chaotic evolution of contemporary societies. In the third space knowledge is no longer enclosed, padlocked like a treasure. It
pervades everything, is distributed, mediatized, spreads
innovation wherever it is found. Technoscience, the can-
cerous body of collective knowledge, metastasizes anar-
chically. Knowledge is no longer a static pyramid; it
swells and travels along a vast mobile network of labora-
tories, research centers, libraries, databanks, people, tech-

nical processes, media, recording and measuring devices, a network that continues to expand among humans and nonhumans, joining molecules and social groups, electrons and institutions.[3]

The eighteenth century of Robinson Crusoe and the *Encyclopédie* marks the end of an era in which a single human being was able to comprehend the totality of knowledge. The encyclopedia of the commodity space is no longer based on the memory of the living body, nor on the Book, nor on any closed system at all. It circulates in a space of translations and referrals, whose first philosopher, as Michel Serres has shown, was Leibnitz.[4] Diderot and d'Alembert have now abandoned the architectonic diagram, the well-ordered hierarchy, since the *Encyclopédie* is now arranged in alphabetical disorder. A hypertext, organized according to its network of internal links. The encyclopedic library pushes the Book aside. And the library continues to expand, overflow, attempts to find its way through file cards and indexes. Citations and references describe a network of readings, travels through knowledge that dissolve ancient borders. Soon, scientific journals will grow in number, drowning us in seas of articles, which will in turn supply innumerable databanks. How will we be able to classify and organize scientific and technical information, as well as museum images, audiovisual archives, and the rising tide of information in general? Our clumsy efforts to make use of the rigid corpses of aging disciplines are doomed to failure, since knowledge exists only at the shifting margins, the crossroads, in interference, when everything is a question of import–export.

Technoscience doesn't produce only a knowledge of chaos and fractals, it also fabricates a chaotic, fragmented knowledge. Once again, things circulate and are distributed: colloquia, conferences, symposia, visiting professors, foreign students, multidisciplinary forums, mediatization, wavelength wars, communiqués—cacophony. We believe that communication consists in sending and receiving messages. Unfortunately, therefore, the more we "communicate," the less we understand. The commodity space condemns us to background noise, agitated crowds, large numbers. Scientometry continues to generate its statistics on publications and patent applications.

Paradigms are no longer popular. There is no longer time to form territories; it's no longer the most pressing problem. For now we must distribute, circulate, network. There is no time for theorizing; we need to model, simulate, operate. The televised premier, the simulation, the computer graphics "demo" are to theory what the audiovisual clip is to the conventional novel. Science and the media echo one another, interpenetrate, help inflate the sphere of untethered signs. Computer-mediated scenarios about the origin of the universe, nuclear war, the hole in the ozone layer, or global economy multiply. They are the result of a fast-moving computer-aided imagination[5] that leaves the figures of ordinary epistemology in the dust. Experiment and theory, though nostalgic, stare at one another like porcelain bookends lying among the territory's abandoned paradigms.

The Knowledge Space: Cosmopedia

In the fourth space knowledge is immanent in the collective intellect. This does not, however, represent a return to the earth, for although knowledge is immanent in its subject, its immanence is radicalized and cuts across both territorial space and deterritorialization. It is an immanence that is without unity or a code.

The knowledge of a thinking community is no longer a shared knowledge for it is now impossible for a single human being, or even a group of people, to master all knowledge, all skills. It is a fundamentally collective knowledge, impossible to gather together into a single creature. All the knowledge of the collective intellect expresses singular becomings, and these becomings compose worlds.

The intelligent community is no longer the closed, cyclical subject of earth, reunited by blood lines or the transmission of stories from generation to generation. As subject, it is open to other members, other collectives, new skills, and is continuously in the process of composing and decomposing itself as it wanders through the knowledge space. In this fourth space the subject of knowledge is shaped by its encyclopedia. Because its knowledge is a knowledge of life, a living knowledge, it is what it knows. And it is precisely because of this reciprocal construction of identity and knowledge that we refer to this fourth, anthropological space as the knowledge space. The philosophical tradition begun by Kant abandoned ontology, the question of being, and concentrated on epistemology, the theory of knowledge. In contrast to Kant's critique, the perspective opened by the collective intellect shows that epistemology ultimately leads us back to ontology: There are as many qualities of being as there are ways of knowing.

Encyclopedia signifies "circle of knowledge," the cyclical interaction of knowledge and instruction. The circle is a figure that, although closed and often seen to represent the infinite, exists only in a single dimension. It is a line. This figure accurately reflects a knowledge that is most conveniently expressed in the form of text, for text is physically linear (even if its semantic structure is much more complex). The process of joining the line (making it circular) connotes the operation of indefinite referral characteristic of the encyclopedia. Although the term existed in antiquity, the encyclopedia, strictly speaking, is the typical form of the totality of knowledge in the commodity space.

Michel Authier and I refer to the new organization of knowledge in this fourth space as the *cosmopedia*.[6] It is based largely on the possibilities made accessible to us through computer technology for the representation and dynamic management of knowledge. Why do we say that the sum of knowledge is now organized by the cosmos and not the circle? Because instead of a one-dimensional text or even a hypertext network, we now have a dynamic and interactive multidimensional representational space. Instead of the conjunction of image and text, characteristic of the encyclopedia, the cosmopedia combines a large number of different types of expression: static images, video, sound, interactive simulation, interactive maps, expert systems, dynamic ideographs, virtual reality, artificial life, etc. At its extreme the cosmopedia contains as many semiotics and types of representation as exist in the world itself. The cosmopedia multiplies nondiscursive utterances.

In keeping with the world and living thought, the landscapes and borders of the cosmopedia are in motion, with regions of varying stability. Its maps are continuously being updated. As is true of the world itself, discourse alone is insufficient for an investigation of meaning. Our explorations are conducted using our sensibilities, along pathways and adjoining areas that are full of meaning. Cosmopedic knowledge brings us closer to the lived world. By this, I mean that our relationship to knowledge-filled representations can simulate our aesthetic relationship to the world by integrating elements of its sensibility, imagery, even its imaginary dimension. While the cosmopedia can be interpreted metaphorically as the ideal figure of knowledge in the fourth space, from a technical point of view, collective intellects are effectively capable of constructing their own cosmopedia.

The characteristic principle of the cosmopedia, and that which makes it worthwhile, is its non-separation. For collective intellects, knowledge is a continuum, a large patchwork quilt in which each point can be folded over on any other. The cosmopedia dematerializes the boundaries between different types of knowledge. It dissolves the differences between specializations, as separate zones of power, and leaves behind regions with fluid borders, structured by concepts of variable significance and objects that are continuously being redefined. In place of the fixed organization of knowledge into discrete and hierarchical disciplines (typical of territorial space)—or the chaotic fragmentation of information and data (typical of the commodity space)—there now exists an unbroken, dynamic topology.

The members of a thinking community search, inscribe, connect, consult, explore. Their collective knowledge is materialized in an immense multidimensional electronic image, perpetually metamorphosing, bustling with the rhythm of quasi-animate inventions and discoveries.

Not only does the cosmopedia make available to the collective intellect all of the pertinent knowledge available to it at a given moment, but it also serves as a site of collective discussion, negotiation, and development. A pluralistic image of knowledge, the cosmopedia is the mediating fabric between the collective intellect and its world, between the collective intellect and itself. Knowledge is no longer separated from the concrete realizations that give it meaning, nor from the activities and practices that engender knowledge and that knowledge modifies in turn. Depending on the zones of use and paths of exploration, hierarchies between users and designers, authors and readers, are inverted. A person who decides to learn about a topic in biochemistry or the history of art will be capable of supplying new information about a given sector of electronics or infant care, one in which he or she happens to specialize. In the cosmopedia all reading is writing. The cosmopedia is a relativistic space, which curves when we read or write in it. Inscription is a form of surgery (cutting, sewing, grafting, discontinuous operations in general). Consultation, however, is a way of massaging or folding space (inflection, continuous operations). Unanswered questions will create tension within cosmopedic space, indicating regions where invention and innovation are required.

In contrast to the expanding complexity that we attempt to organize through transcendence or distribute within increasingly inextricable networks, the cosmopedia provides a new kind of simplicity. Not some mutilating simplification applied violently from without, but an essential simplicity that results from the principle of organization inherent in the knowledge space. Beyond the world of chaos and large numbers, beyond the incantations to complexity, simplicity is born of implication.

The continuous space of proximities that shapes the cosmopedia implies, in its dynamic structure, the relationships, links, and connections among utterances. The situation, context, assumptions, and conclusions of a proposition no longer need to be made explicit in speech since they are implied in the moving form of the image. The context, references, conceptual beliefs, or practices of each proposition no longer need to be remembered since they are always already there. The presence of new propositions is made known (to other propositions or entire swaths of the cosmopedia) through their relative position, proximity, color, and brightness.

The collective intellect shapes, molds, smooths, and sculpts the image of its knowledge and its world rather than translating it discursively. Simplification is the result of the considerable reduction of the importance of text in the exposition of knowledge, which results from the inclusion of relational information in the very structure of "cosmopedic space." The transition from the portolan to the map serves as an analogy. Here a large number of tales and stories were summarized in a single image that contained all the information needed for navigation. But two fundamental issues distinguish the invention of the cosmopedia from the use of navigational maps. In the first place, it is the knowledge space itself that is dynamically mapped and not simply a part of the universe of reference for a given community. Second, the simplification does not result from the projection of a system of transcendent coordinates but of self-organization in the plane of immanence.

If we were to ask *who* implies the relations among utterances within the space of the cosmopedia, the answer would be the collective intellect itself, its navigations, the trail of its inscriptions, the footprints left on the plane of immanence of its knowledge. The relations between cosmopedic utterances are implied in the structure of the large, multidimensional image only because the living members of the collective intellect are as well. It is they who secrete and weave and sew and fold the knowledge space from within.

Once they plunge into the cosmopedia, space reorganizes around them, depending on their history, their interests, their questions, their previous utterances. They are surrounded by everything that concerns them, which arranges itself within their reach. The things they are least concerned with move off into the distance. Distance itself is subjective. It is through the process of implication that we filter the large numbers typical of the commodity space. It is through the simplicity of our immersion that we escape its complexity, its labyrinthine networks.

Once within, the member of the collective intellect swims around (navigates, consults, questions, inscribes, etc.), then leaves. Memory of the digital waters: His swimming has modified the structure of shared space as well as the shape and position of its image in the cosmopedia (his personal navigator). It is the same for everyone, each time they dive into the cosmopedia. Together, they organize the space, define, evaluate, color, heat, or cool it. Each one helps build and order a space of shared signification by diving in, swimming around, and simply living in it.

The Philosophy of Implication

Our focus has been on the collective intellect and its world rather than culture or nature. The world of the collective intellect is both its universe of reference, its "culture" (if we insist on using the term), its perception of self, and its effective identity in the knowledge space: inseparable aspects of the same fluid reality in a process of self-organization.

For contemporary epistemology, whose association with territorial space I indicated earlier, the subject constructs its object of knowledge. This may be a transcendental subject, which imposes upon its object the *a priori* forms and categories through which it apprehends (time, space, causality, etc.). It might also be a scientific subject that subjugates an object to its measurements, concepts, and theories. Historical transcendentals have been proposed, abstract subjects that emerge from cultural configurations, which elaborate and perceive their objects by means of languages, recording and communications technologies, institutions, organizational, imaginational, or symbolic forms. Empiricism, on the other hand, has described objects as imprinting themselves on the subject, an intelligence shaped by its experiences. Between these two positions (empirical subject imprinted by reality or transcendental subject) several intermediary positions have developed: innate structures triggered by experience, the complicated system of accommodation and assimilation, the dialectic of interaction. Although all these approaches to the relationship between subject and object of knowledge have their own validity, none of them corresponds to the situation that exists in the knowledge space. For although they begin with the subject, or the object, or their interaction, both terms are initially understood as external entities.

In the knowledge space the object constructs the subject. Once again, the object here is the perpetual renewal of the becoming of the collective intellect and its world. It is as if the subject were fabricated by the subject.[7] As for the world, it is no longer an "objective" world, but the world of the collective intellect, the world that *thinks in it*.[8] In thinking itself, the world of the collective intellect invents itself as collective intellect. This world, which is always in a state of formation, creates the becoming-identity of the collective intellect. The subject implies the object. The diaphanous fabric of the collective intellect envelops its thinking world.

By the same movement objects in the knowledge space are constituted by their collective subjects. Here, the subject no longer constructs its object transcendentally, from above, as something external, capturing it in a filter or on a grid, but through implication. The collective intellect aggregates its practices, hopes, interests, and negotiations, deposits its living outbursts of energy, sediments its subjective becomings, concretizes its affects, and, having done so, secretes its world. In the knowledge space, however, the object of knowledge is precisely the cognitive dynamic that carries with it the reproduction of the intelligent collective. It is as if the object produced itself. The object implies subjectivities that are piled up, pressed, massaged, and continuously added to.[9] Consequently, *to know* something, *to be implicated in an object*, means that we give it existence. The collective intellect, by its very nature, is always in the process of knowing and, therefore, of fabricating, its world, of producing being. In the knowledge space objects and subjects are always already implied in one another. Like obverse and reverse, its world, from its inception, is the opposite side of the collective intellect.

The philosophy of implication is not simply a contemporary avatar of Hegelian absolute knowledge, for the forms of knowledge characteristic of the earth, the territory, and the commodity space remain. They are not absorbed, suppressed, phased out, or preserved in the knowledge space, as a Hegelian philosophy might imply, but subsist as integral elements of their own space. Nor does the knowledge space incorporate the internal dialectic of a single and unique molar mega-subject of absolute knowledge—Mind, God, or the Philosopher—capable of transforming all contingency into reason, of translating totality. On the contrary, an indefinite variety of collective intellects smooth out, extend, and constitute the knowledge space, each unfolding its own world, so that this space is always exposed to alterity, to other spaces, to an indeterminate future. It lives by reason of this creation and heterogenesis.[10] It is essentially plural, and in the process of dynamic pluralization. Far from being governed by an organic and necessary progression, quasi-eternal, it is animated by the growth of singularities awaiting creation. In the knowledge space, thoughts are worlds in an emergent state. Hegel described the molar becoming of a total subject. Collective intellects encourage processes of molecular, bifurcating subjectivization. In the Hegelian system becoming is the self-movement of the concept. In the context of collective intellects, the self-movement of becoming expresses itself through ontological and conceptual productivity.

In spite of these differences between Hegelian philosophy and collective intellects, an essential affinity remains: the attempt to reconcile thinking and being, even if it involves, for the collective intellect, an unbounded variety of qualities of being and modalities of thought. Thinking and being, identity and knowledge. The collective intellect and its world are not satisfied merely to coincide, they are engaged in an uninterrupted process of pluralization and heterogenesis.

15

The Relationship between the Spaces

Toward a Political Philosophy

Successive Eternities

In the preceding chapters, I tried to indicate the specific characteristics of the four spaces and their respective differences. I would now like to reexamine some of their shared features, their structure, and the ways in which they might coexist with one another.

These spaces can't be classified as eras, or ages, or epochs for the simple reason that they can't be substituted for one another but coexist. And yet, as structuring and autonomous spaces, they appear in succession. Since a space becomes irreversible once it has been deployed, it can't be eliminated by succeeding spaces. This results in a kind of anthropological geology in which the spaces function as layers. While the metaphor is useful, it shouldn't be overworked.

These layers can be characterized independently of the time of their appearance and on the sole basis of the quality of being they radiate, the sign that identifies them, or the principle that creates them. The symbol of the earth might be the sphere, enclosed, unique, and replete. Its principle is "be world," our world, a cosmos. The insignia of the territory could be the pyramid and its organizing principle, transcendence. The network, or circuit, could be the icon of the commodity space, and deterritorialization its principle. As to the knowledge space, its emblem is the reunified tree of knowledge and life, and its principle radical immanence. The figures and principles of the anthropological spaces have no relationship to chronology; they are defined by the modes of existence they engender. Their forms have been, are, and will be present at all times, in varying degrees, and in varying proportions. Collective intellects, in particular, the subjects of the fourth space, have already expressed themselves on numerous occasions, with varying degrees of consistency and intensity.

Each situation, each social dynamic can be understood as a configuration of spaces, a combination of figures, a particular arrangement of principles. In this sense the anthropological spaces are eternal. And yet, once again, we find a succession of spaces if we consider, not their characteristic elements, figures, and principles, but their deployment as irreversible and autonomous anthropological spaces, as fundamental organizers of the great periods of human adventure. Conceptual, outside time, but temporalizing, the anthropological spaces are produced and maintained by the activities of living human beings. These are the acts of men and women, their thoughts and relationships; they actualize a given space, extend it, suffuse it with reality.

Although they succeed one another, none of the spaces has ever been made obsolete. Each of them is always at work, awaiting a more intense reactivation. We can use an analogy from our subjective experience: Time doesn't really pass; affective environments, existential configurations are put in reserve, stored in memory. And since they are always operational, they are available at all times. Everything is always present. We live in keeping with the footprints on the earth, the enclosed spaces and gates of territory, the networks of commodity, the interior spaces of knowledge. We return to the immemoriality of the earth, exist within the slowness and deferral of the territory, accelerate to the real-time of commodity space, and grow in tune to the subjective temporalities of intelligent collectives.

Spaces and Strata

Earlier, I defined a stratum as a way of analyzing reality, a way of dividing up its living, cosmopolitan tissue, which connects all the domains of mankind.[1] The anthropological spaces should not be confused with such strata, however. For example, the earth is not ecology, the territory can't be identified with the city, commodity space can't be superimposed on economy, all forms of knowledge are not contained in the knowledge space, and the knowledge space itself is not limited to cognitive activities in the strict sense of the word. Politics, economy, and knowledge are strata, analytical categories, and in no way living spaces of signification, perpetually created and re-created by the activities of human beings.

In preceding chapters I have discussed the ways in which the different spaces redefine and alter knowledge. The stratum of knowledge is thus perfectly orthogonal to the other spaces. Each stratum assumes a different figure depending on which space it intersects. I have devoted more attention to this phenomenon in the case of knowledge, because the subject of this book is collective intelligence. There now remains the question of how the other strata are regenerated by the spaces they intersect. I will simply outline the political and economic relationships among the four spaces, my principal objective being to distinguish these strata from the spaces they appear to correspond with.

With respect to the stratum of economy, it is important to realize that there are other relationships to wealth than those characteristic of the commodity space. The economy of the earth is organized around gift-giving, spending, potlatch, communal sharing, and pure rapaciousness. In territorial space the economy is administered and managed for the long term. It is based on the regular collection of tributes, taxes, tithes, and rent. The economy is obviously capitalist in the commodity space, capitalist and thus exchange- and production-based, whereas preceding spaces are characterized by gift-giving, plunder, or rent. The knowledge space may give rise to an economy of knowledge,[2] some of whose underlying principles were discussed earlier.

It is important not to confuse the manifestations of economy in each anthropological space—the qualitative mutations they undergo—with the projection of capital upon these spaces. For example, the projection of capital on the knowledge space is obviously labor, which is no more than a degradation or flattening of the life of the intelligent collective. The projection of capital in territorial space gives rise to real-estate, an object of speculation wrenched from the economic logic that characterizes the second space. The projection of capital on earth is the natural resource, which denies and expropriates the ancestral earth, the inhabited earth, and which obviously has little connection with gift-giving or sharing, even though it constitutes a kind of monstrous return, on a large scale, of Paleolithic predation. This notion of the projection of one space on another will be treated at greater length in the remainder of the chapter.

Despotism, bureaucracy, and representation are the major features of politics only within territorial space. Its form is constructed in opposition to the tribal or clan communitarianism typical of the earth. Today, it is combined with thermodynamic democracy, the original form of politics in the commodity space. Thermodynamic democracy channels a large number of collective problems, ideas, and practices into simple binary choices: yes or no, right or left, republican or democrat, conservative or progressive, etc. Qualitative diversity and living, organic composition are represented by quantities that are distributed between opposite poles. Like the hot and cold of thermodynamics, we end up with a lack of differentiation: tepidness, statistical averages. Everything ends up looking the same. As I have shown in Chapter 4, the canonical form of politics in the knowledge space is a kind of direct, computer-assisted democracy no longer based on the representation of statistical majorities but on the self-organization of intelligent collectives, in which minorities[3] have an opportunity to experiment and take initiatives. All these forms of economy, all such types of political organization, coexist today in changing configurations that vary from situation to situation.

Toward a Generalized Human Ecology

We are all aware of the size and complexity of the problems currently facing mankind and the powerful changes affecting our societies. They require that we reshape the economic and political categories that were forged at a different time to resolve different problems. It is for this reason that I am not proposing a political program but a conceptual framework, a philosophical approach, which I hope will supply effective instruments for comprehending and resolving society's organizational difficulties.

The following outline of political philosophy reflects a form of human ecology, the art of establishing harmonious relations among the four anthropological spaces. Such a political philosophy implies the creation of a fourth space, the site of a democracy of initiative and direct experimentation, employing new technical and social instruments of collective expression that promote, and do not repress, singularity. This fourth space cannot be instituted by fiat; it will spread and grow with the life of the collective intellects that animate it.

Yet, in spite of the importance of an autonomous fourth space, my political philosophy is strongly opposed to any notion of abolishing or suppressing any other anthropological space. More than ever before, we are cognizant of where the desire to start from a tabula rasa, the urge to violently overthrow existing systems, can lead: to a return of the abolished that is even more pernicious, more monstrous than its earlier incarnations. Starting over, making revolutions, inaugurating a new era, are all projects that belong to the mythology of the territory and are thus incapable of orienting the political philosophy promoted here.

As mentioned previously the first three anthropological spaces are irreversible and, to some extent, eternal. We must now turn our attention to their relationships (with one another and the virtual space of knowledge); not simply their effective relationships, which we can observe on a daily basis, but also their desirable relationships. A political philosophy that takes itself seriously must do more than merely analyze or dissect a situation. It must also point the way toward a positive outcome of some sort.

Conditions and Constraints

I will postulate that in principle no space can or should reduce, assimilate, or destroy any of the other spaces. This statement is based on the fact that the anthropological spaces depend on one another. They are mutually conditioned and, in particular, the full, entire, and autonomous existence of inferior or anterior spaces acts as a severe constraint on the development of superior or posterior spaces.

No government can exist for long if it destroys its subjects' relationship to the cosmos, either practically or symbolically. No commodity economy can experience long-term development if it weakens the biological and imaginary roots that unite mankind to the world. This is merely the result of an elementary calculus of existence: There are ecological conditions (both in the environmental and anthropological sense) that must be fulfilled to achieve either state power or economic prosperity. The knowledge space itself is obviously nourished by the world of archetypes, cosmic figures, felt intuitions, and narratives of the earth.

The active presence of the territory, in its turn, conditions the growth of the knowledge and commodity spaces. Without a stable and respected state of law, without the possibility of ensuring that contracts are fulfilled, without effective and impartial public services, the economy cannot grow. Without schools, educational institutions, museums, libraries, etc., without public research institutions or communications infrastructures, can the knowledge space exist at all?

The commodity space conditions the knowledge space in the sense that the collective intellect, if it wishes to endure, must respect certain management rules, certain basic economic constraints. Collective intellects will benefit from the real-time technologies that emerged within the commodity space. Their immanence *follows* deterritorialization and is established in the presence of a transcendence that has already been weakened, pluralized, and distributed by the commodity space. The collective intellect may require the existence of a powerful buffer space between itself and the territory.

Semiotic Efficiency and "Inferior" Spaces

I will try to show how, in the case of signification, the spaces below condition those above. In a previous chapter I described the semiotic that corresponds to the four anthropological spaces. But we haven't yet examined how signification functions in actual situations, situations in which humanity simultaneously inhabits all four spaces.

No semiotic operates without precedent. The earth (here, the consubstantiality of signs, beings, and things) is and remains the soil of meaning. Although the territory may impose a form of semiotic separation, the transcendent domination of the symbolic or the signifier, meaning could never be achieved if the earth were not active at all times. Languages can be codified as languages, and not reduced to codes, only if the earth continues to animate them from below. To a child, words are still real attributes of things. In some sense, speech may be impossible without the indistinction between sign and object, without breath, the living presence that establishes a continuity between sign and being. It is against the permanent background of such continuity that the territory can write, separate, divide, codify, and legislate meaning.

From the point of view of emission, the spectacular commodity machines must be fed with difference on earth and in territorial space, so they can introduce variety into the mediasphere, always in the process of thermodynamic indifferentiation. On the receiving end, as divorced as it may be from any need for genuine representation or coherent discourse, the spectacle is forced to build upon the substrate of the dictionary and grammar. It is because television viewers (or newspaper readers or the radio-listening audience) associate perceived signs with territoriality, with things, that they pay attention to the media, even though its own logic is very different from that of the representation. And media is even more dependent on the earth than it is on commodity space: It only makes sense because the public is able to integrate it with its cosmos. Cut off from the living context of a song in a foreign language heard on the radio, I construct around it an entire world of affective relations. Without the depth of meaning associated with the two preceding spaces, the spectacle would be unendurable. Newspapers give me the illusion of supplying something I can decrypt only by referring me to a territorial semiotic of the representation and truth of writing. We are affected by music or film only because they awaken the terrestrial being in us.

If the viral semiotic of propagation, unlimited circulation, and pure distribution were to eliminate all the others, then, and only then, would media domination become a cultural catastrophe. We can demonstrate, in fact, that the signifying productivity of collective intellects is rooted in the existential plenitude of the earth, that it implies mastery of the codes of the territory, and that it makes use of the techniques of proliferation and irradiation characteristic of commodity space.

Causality without Contact

Each new space rests on the preceding spaces. No anthropological space can destroy those that are below it without the risk of destroying itself. Having clarified this, however, we are still far from a full analysis of the relationships among the spaces. At this point the anthropological spaces are no more than a succession of strata, a hierarchy of conditions and constraints, a system of infrastructures and superstructures. Unfortunately, this simplistic view of the spaces does not provide an accurate picture of the nature of their interrelations.

Concrete situations and beings are immersed in several anthropological frequencies at once. Because no real being can subsist in a single ether or without the existence of communication among the spaces, we can say that the anthropological spaces depend on one another. Each space, however, is fully autonomous and neither perceives nor modifies the others, except in accordance with its own specific principles, through reference to itself, by seeing the reflection of its own figure everywhere around it.

Concrete beings and entities, cosmopolitan machines traverse the four velocities, but none of them can have a direct effect on the other, or even touch it. The anthropological spaces are related to one another but only in terms of a causality without contact. For example, everything that a collective intellect perceives will be integrated in its world, evaluated according to its criteria, subjugated to its own temporality, metamorphosed through appropriation, so that the entity under consideration will no longer belong to another anthropological environment. The commodity space is incapable of apprehending the knowledge space as such. We can't buy or distribute the thought of a collective intellect. We can only commercialize its projection on the commodity space, propagate unattached signs in the media market. But in doing so the reciprocal interaction of the collective intellect and its world is dissolved. Thought is no longer involved. Just as the objects that King Midas touched were changed into gold, the commodity space deterritorializes everything it distributes throughout its networks. Indeed its situation is worse than that of the king of Lydia for it can never touch anything but itself.

The same is true of territorial space. As soon as it takes hold of an entity from another space, it fixes it according to a set of transcendent coordinates, encodes it, inscribes it within a tradition. The territory is destined to ignore any other time but historical time, any space that is not hollowed out by foundations or separated by borders. Thus no anthropological frequency has a direct effect on the others since it is always involved with itself. There are no actors, only spaces. How then can we conceive of causality without contact, reciprocal modulations among worlds that ignore one another, the forces of anthropological gravitation?

It is as if two currents, one rising, the other falling, ordered the relations among the spaces. From bottom to top, the slower, deeper spaces are attracted by the faster spaces above them. The lower spaces are moved or disturbed by the upper spaces, through their desire. Thus, in one sense, the earth desires a territorial space that fascinates and subjugates it; the commodity space desires the knowledge that escapes it and causes it to circulate.

Inversely, going from top to bottom on our anthropological Jacob's ladder, the uppermost spaces open themselves up to the lower spaces, nourish them in turn, without perceiving them, by remaining always inside their own substance. Yet "the path from top to bottom and from bottom to top are one and the same."[4] For in one sense, the earth is like an ocean that gathers together all sentiments into which all the rivers of the mind flow together. If we overturn the system, however, the knowledge space in its turn becomes an ocean, absorbing the

impalpable wave of living knowledge. Collective intellects, hidden stars illuminated by their own light, swell
the tides of meaning.

Release and Desire

Although only virtual, the knowledge space, the
space of collective imagination and creation, nourishes all
the other anthropological frequencies. It is like the divine
effusion, which in medieval cosmology flowed and overflowed from intelligence to intelligence, from sphere to
sphere, until it reached the sublunary world.

Each new anthropological space encloses the previous space, imposes its signification, its direction, its
velocity on it. The history of empire obscures the immemorial cycles of the earth, the accelerated rhythms of
industry overturn the slow pace of peasant societies. The
subjective temporalities of collective intellects are in turn
capable of modifying the real time of commodity networks. The society of the spectacle is that intermediary
moment during which the informational sphere has already acquired some consistency without becoming fully
autonomous with respect to commodity space. Try to
imagine the power of digital and media technologies in
the service of the collective imagination, used for the
continuous production of subjectivity, the invention of
new qualities of being.

There are people in the commodity or territorial
spaces who fear the establishment of a space of collective
invention. They are unaware that competition is impossible among the spaces. By preventing the knowledge
space from becoming autonomous, they deprive the circuits of commodity space and the territory of an extraordinary source of energy. The knowledge space will
nourish the commodity space to the extent that it can free

itself of it. The sooner intelligent collectives escape the territory, the better off governments and institutions will be, once freed of their stifling castes, bureaucracies, and history. The short-sighted individuals who resist the autonomy of the collective intellect, incapable of deploying their being throughout several spaces, are like the men of fable: By killing the goose that lays the golden egg, they impoverish themselves.

If the knowledge space were to become irreversible and guide the commodity space in accordance with the causality-without-contact I have attempted to describe, then, perhaps, the external and violent velocities of networks, interaction, instantaneous adaptation, and senseless flux, rather than being experienced as imposed and destructive necessities, would become the byproducts of real-time composition, the manifestation of internal necessity. Just as the urban is the externalization of commodity circuits within territorial space, the networks of commodity space may in turn concretize the subjective temporalities of collective intellects or imagining collectives.

Each space thinks and desires the others on its own terms, according to its own figures. In this way the earth loves commodity space: It creates cargo cults, makes magic out of technology, constructs an industrial and mediatized cosmos, a strange global habitation. Of course, industry continues to humiliate the earth, to deny it, for it has never really encountered it. But isn't it commodity space that mediatizes and unifies the planet, provides it with a consciousness, a schizoid consciousness, fragmented, intermittent, and mad?

The earth also desires the knowledge space, for it alone of all the spaces understands and recognizes the earth. Its subjective nomadism leads it to the originating, though impalpable, nomadism of its ancestors on earth. Knowledge can never leave the earth in peace, however.

It continues to invent unbound worlds whereas the earth prefers to reconstruct a unique and enclosed cosmos. The lovesick earth is overwhelmed, heterogenized, fecundated, mutated by the knowledge space.

Even the territory desires the commodity space that deterritorializes it. Governments forget history and aimlessly follow the rhythms of the media. Cities and regions battle with one another to harbor the industries they live off, to attract the hubs of rail and air networks, to find a place among the great currents of exchange. States are no longer merely dependent on economy and the media. They fear the loss of intellectual skills[5] or worse, the flight of entire peoples, deprived of collective intelligence and imagination. The exodus caused the pharaoh to tremble more than the approach of a thousand enemy armies, destined to bring other pharaohs in their wake. The territory tends toward the knowledge space, though maladroitly: technopolises, universities, silicon valleys, transparency....

The commodity space in turn suspects that an as yet unknown economy is being developed in another space, one it will never be able to comprehend, an economy of knowledge, independent of the sphere of capital and no longer subjugated to the rule of money, but driven by other principles. Astonished, it discovers that this mental and subjective economy—the life of intelligent collectives—desires and engenders the commodity space, serves as its invisible and intangible engine. The commodity space, in turn, desires the knowledge space.

The relationships just described are harmonious. The spaces above spread their substance throughout the spaces below them. The lower spaces desire the spaces above. The ideal is achieved when the knowledge space assumes autonomy, becomes irreversible, and imagining collectives polarize anthropological gravitations, sentiments, and circulation, thus establishing the most fluid and unfettered system of relations possible.

The uncrowned sovereignty of knowledge arises from the fact that it is always desirable, intangible, mobile, living, fecund, multiple, never feared. The word "knowledge" here doesn't imply that power will be given to intellectuals, researchers, experts. I am referring to an ideal that is radically different from the Platonic city state, directed by its philosopher caste. Plato was describing a territorial utopia, an earth immobilized by a philosophical sky. Our knowledge space, on the contrary, is the siteless site in which collective intellects wander, in which qualities of being are invented, a place beyond deterritorialization, an infinite plane of immanence, open and unbounded. It does not define some ideal of static perfection but the principle of self-organization, the continuous self-invention of human communities and their worlds.

The Four Cardinal Points

Having indicated the form of the harmonious relationships among the spaces, we can now more clearly identify the cacophonous relationships among them, those that almost inevitably lead to evil and servitude, stifle ontological invention, and destroy life. The most harmful relationships are those that occur when the bottommost spaces attempt to violently control those above.

Evil arises from the earth's desire to control territorial space, when the tribes destroy one another in their attempt to gain possession of the state, when a clan chief becomes the head of government. It is the type of misfortune that plagues the countries of the *South* today, torn apart by civil wars, dictatorships, and famine. Evil arises from earth's desire to subjugate itself to the commodity space, when industry and commerce are controlled by clans, when predation is substituted for exchange. Banditry and organized crime reign in a different South. When the earth is involved in driving the knowledge space, it gives rise to the New Age, ecological fundamentalism, militant irrationalism.

I refer to the territory's desire to control the commodity and knowledge spaces as the evil from the *East*, in memory of the great glaciation that marked Europe during the twentieth century. The absolute domination of the commodity space by the territory gives rise to a controlled economy, or planned poverty. Isn't totalitarianism the attempt to make the spectacle subservient to the territory? Subject to the territory's will to control, the knowledge space is unable to survive in an embryonic state, but is immediately destroyed or condemned to a dangerous kind of clandestine existence. Territorial space's attempt to control the spaces above it is not simply a thing of the past. The East is all around us. In large (and small) companies, state-like bureaucracies interfere with economic initiatives, administrative routine stifles invention, authoritarian control and separation prevent collective intelligence from spreading. The bureaucratic and institutional form assumed by universities, research centers, and schools is no longer favorable to the growth of collective intellects. There is also an East of education and official science.

To the *North* the commodity space claims to govern the knowledge space. This is the disease of our wealthiest nations. In the society of the spectacle, thought is buried in the world of media and advertising. In place of collective intellects, the North can only provide technoscience, finance, and media, the madness of large numbers and velocity, unrestricted deterritorialization, external, violent, and without subjectivity. This North has spread throughout the world.

Where does our current sense of disorientation come from? The discontented of the North, those who have been wounded by deterritorialization, can find no other outlet than through an appeal to transcendence, a return to the hierarchies, traditions, histories, and values of the territory. Leaving this North, we can turn eastward, but we face an East that continues to perpetuate and disseminate itself. Those who are disappointed with the East will head North, oscillating between the state and capital, as if they were the only two poles of existence in the world. Others will turn to the South, where they anticipate the terrifying domination exerted by the earth. What about the remaining direction? Few people see it because it is deserted, because its explorers have left no trace of their passage.

This text is an attempt to point a way toward the *West*, by indicating the empty, unexplored ocean, the great discoveries ahead. The West is a convocation to depart, a silent appeal to discover a new space.

Emptiness and Plenitude

Now that we have examined the velocities of the anthropological spaces, their relationships, their articulations, we need to reestablish the essential continuity of the human. How can we travel from one space to another? Where can the individuals and collectives who cross the four spaces reunite, construct a quilt in which there are no stitches among the figures it comprises? In spite of the foreignness of the principles that divide them, can some common ground be found?

In territorial space, when a border must be crossed, we erect a customs post. The customs agent expands his booth, fortifies his barriers, complicates procedures, until his customs booth has become another territory, an additional enclosure. Other customs posts will have to be constructed, other intermediaries needed to cross the new territory. By establishing ports, canals, locks, gates, the *parasite* pretends to promote circulation but ends up creating its own territory. Other intermediaries are always required. This is the infernal logic of parasitic inflation.[6] In this manner the labyrinth is constructed. The Minoan palace exists at the dawn of history and the state, an intricate arrangement of angles, redans, booths, corridors, and rooms, the barred and inescapable space of the territory. The minotaur, the sacrificial and devouring face of transcendence, stands in the center of every territorial space.

In contrast to the character of the customs agent, the parasite, stands the ferryman. The ferryman helps smuggle us across borders, he leaves no trace behind, builds no new walls under the pretext of erecting a door. We need a ferryman, not only between the fortifications, enclosed spaces, and internal institutions of the territory, but also to escape the territorial labyrinth and jump from space to space. Daedalus was only able to escape the labyrinth by entering another space.[7] Where is our ferryman?

Commodities circulate and create circulation, but they also snatch, decontextualize, and violently deterritorialize whatever moves within their networks. Instead of reestablishing continuity, commodity isolates and detaches. The territory encloses things through interiority; commodity condemns them to exteriority.

The knowledge space is not simply a way out of the territorial labyrinth but a bridge between the three previous spaces. It enables the earth, the territory, and commodity space to communicate with another. The ferryman is thought within the individual, the collective intellect among divided individuals. Thought can play this role because, unfolding itself across an infinite plane of immanence, without appropriation or inertia, its essence is to welcome being in all its diversity, allow it to coexist. As it evolves in understanding and creativity, the collective intellect finds that it has nothing to defend or sell. All its efforts are directed toward welcoming, making available, understanding, and reinventing its own conscious becoming. The collective intellect works to enlarge emptiness. Not loss or absence, but Taoist emptiness, the openness and humility that alone give rise to learning and thought.

The earth is eternally full. It is the space of the gift, of gratuitousness, of endless profusion and ever-present meaning. Territorial and commodity space operate through loss. Absence, castration, and deferral crisscross territorial space. In the commodity space, loss takes the form of need. Yet the knowledge space is not always full, like the earth, and lacks nothing, unlike the territorial and commodity spaces. The knowledge space is exposed to the void.

Until now we have looked at the earth as a kind of central core in our anthropological cartography, a base supporting the other spaces. On this map, the knowledge space is represented as a sunless sky, which orients and sets in motion the spaces below it. But on a different map, using a different projection system, the earth is all around mankind, its mass rests along the periphery, it envelops the universe of signification of the human world, a universe of meaning and shape, which slowly differentiates itself, decants itself, from space to space, and comes to rest near the center.

Knowledge thus becomes the nucleus of this new anthropological cosmology. Yet it doesn't concentrate elements that are heavier than itself, any particular kind of atom, it embraces emptiness. The focal point of all gravitation, the collective space of knowledge, invention, and apprenticeship, is the centermost void that drives the entire human universe. Though empty, it straddles the places we will pass through, makes movement possible, establishes the essential continuity among all the anthropological spaces.[8] Genuinely human existence—like all true encounters between individuals—is born, perpetuates itself, and finds its unity in thought. It is suspended in the void.

Epilogue

Voyage to Knossos

Is the prospect of developing a collective intelligence realistic? Is it an impossible utopia or a feasible alternative? Before trying to answer these questions directly, it might be a good idea to explain what I mean by feasible, possible, and real.

The *possible* encompasses systems of noncontradictory facts. While these do not contradict any known physical laws, they do not take present circumstances into account. The *feasible*, on the other hand, represents a much more limited range of items than the possible. Its options incorporate resources that are available here and now, and respect the technical, economic, and social constraints imposed by the situation. Our effective choices or decisions select facts from among the range of the feasible. Action lies at the interface between the feasible and fact. A living and permeable membrane, an active and subtle filter, action transforms feasibility into fact, extends the domain of the effective and transforms the realizable.

But an act can also have an effect on another interface, one that lies upstream of the ontological flux and separates the possible from the feasible. These are the technical, economic, and social conditions separating that which is simply possible from the constraints of feasibility. Technology, in the broadest sense of the word, refers to any event whose effect will be to shift the boundary between the possible and the feasible. The project for collective intelligence enhances technology, not through blind fascination but because it opens up the field of action. Skill and technology are important for two reasons: primarily as products, representing the crystallization and memory of human activity, and then as potential instruments for enhancing our powers of understanding, feeling, acting, and communicating, as an interface between the possible and the feasible. Of course, technology can be used in such a way that it results in an overall decrease in power and greater exclusion rather than sociability. But in this case we squander our human qualities: those belonging to the producers—since we have made use of an intelligence that resides in things for the purpose of destruction—and those belonging to the victims. Failure to exploit technology also involves waste, however. For its inventors will have worked in vain and potential beneficiaries will have been deprived of additional qualities. It is the degree of collective intelligence at work in a given situation that conditions the human value of technology. Everything depends on the collective's ability to enhance the power of practical knowledge and material technology available—an enlargement of the domain of the feasible—through an overall enrichment of the human.

Through their effects, our actions can shift another limit, another interface, that which establishes the division between the impossible and the possible. Critics may claim that this limit can never be altered, that what is impossible has always been and will always be impossible, by definition, otherwise it wouldn't represent true impossibility. Imagine, however, an Aristotelian and rationalist astronomer of the Middle Ages. For him, the Moon and Earth are absolutely distinct spheres of existence. He knows from the laws of science that it is impossible (even though imaginable within a story or fable) for a man, a living and mortal human being, to walk on the Moon. For Galilean science, however, the lunar and sublunar worlds are no longer separated by a radical difference in their nature. A Newtonian astronomer would claim that it is possible for a man to walk on the Moon (even though he doesn't see how this could be feasible). In the twentieth century, technology has made the possible feasible, and NASA's actions have made the feasible fact. But the initial shift, between the impossible and the possible, was a scientific act, related to the sphere of valid representation and systems of explanation. Today, we know it is impossible for a physical body to exceed the speed of light. We know that, within a formal system powerful enough to model arithmetic, we cannot demonstrate both the consistency and completeness of a system using the resources of the system alone (Gödel's theorem). Using limitation theorems and the laws of physics, science skirts the edges of the impossible. In terms of the description and explanation of the universe, science (in the most general sense of established understanding), to the extent that it shifts the limit between the known and the unknown, because it always reveals unknown forms of becoming, enlarges the realm of the possible.

It is important not to confuse the impossible with the unimaginable. I sincerely believe that it is impossible to exceed a velocity of three-hundred-thousand kilometers a second, but I can still write a science-fiction story in which this impossibility would become quite commonplace. The unimaginable and the impossible belong to separate realms. I can neither know nor say anything about that which is unimaginable for me. I can't even determine its limits. I only know that such limits exist. For example, the contemporary world, with its automobiles, airplanes, telephone networks, televisions, computers, electricity, and nuclear power plants, all the existing details of science and technology, all its political and religious customs, its "mentalities," is strictly speaking unimaginable for a citizen of ancient Gaul, or even for a person living in the seventeenth century. It is our cultural equipment as a whole that shifts the interface between the imaginable and the unimaginable. Here, culture is like a toolbox that is made available to our mental powers: science, technology, our understanding of historical and social facts, language, words and images, ideas, thoughts, ways of thinking, all are intellectual tools, each of which provides its own contribution. Among this hardware, the instruments for observing, simulating, and navigating knowledge uniquely enlarge the scope of the imaginable and, in this sense, help improve our choices.

Finally, thought itself shifts the interface between the imaginable and the imagined. Thought is a producer of images, signs, and mental beings, without which no choice or freedom of any kind would even be possible. Thought extends the range of the imagined and multiplies all the other spaces.

Impossible, possible, feasible, and fact are not only arranged in levels, following a linear ontological scale, they are also organized in terms of a reciprocal, orthogonal interaction, which describes an autopoietic spiral of existence. Unimaginable, imaginable, and imagined not only uniquely determine the three steps of a noetic staircase, but a dynamic spiral of the imagination. We can represent these spirals as infinite movements, increas-

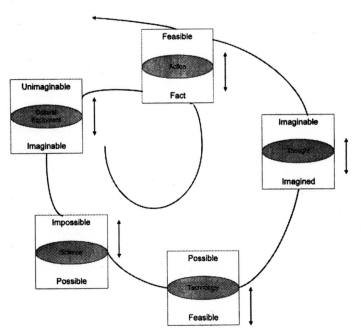

THE CONTINUOUS DRIFT OF THE HUMAN WORLD

ingly detailed interpenetrations of the processes involved. From this point of view, science, technology, action, culture, and thought are not seen as substantial and clearly identified actors but as facets of acts or events. What is an act? That which displaces one of the five borders. If an act is perceived as shifting the boundary between the possible and the impossible, it is part of science. If it acts at the interface between the possible and the feasible, it is part of technology, etc. We never know *a priori* if an event belongs to science or technology, action or culture, or thought, or several of these interfaces at once. We must first examine the limits it shifts. Each time an interface is modified, the others are affected. The spirals of imagination and existence mutually produce one another and collectively compose the great machine that causes the human world to drift along. A machine in which the accretion of the possible, the feasible, facts, the collective imagination, and personal thought are endlessly repeated, an anonymous and singular machine, a floating machine without an infrastructure, without a final causal soil, without foundation: the emergence and turbulence of human reality.

Thus, asking if the idea of collective intelligence is utopian or realistic doesn't really make much sense. The process has begun and we do not yet know, within the context of its overall movement, what limits it will shift or how far it will shift them. Its ultimate finality will be to place the reins of the great ontological and noetic machine in the hands of the human species considered as hypercortex. No attempt has been made in this book to describe a perfect society, devoid of becoming. Collective intelligence is a utopia of the unstable and the multiple. It responds to an ethics of the best rather than a morality of the good. Static, definitive, decontextualized, the good is imposed *a priori*, on top of any existing situation, whereas the best (the best possible) is situated, relative, dynamic, and provisional. The good doesn't change; the best is different wherever it is found. The good is opposed to evil; it is exclusionary. The best, however, includes evil since, logically equivalent to the lesser evil, it is satisfied with minimizing it.

Does the suggestion of a utopia of collective intelligence lead to the myth of progress, an advance toward a future that is always better? No, for the idea of linear progress assumes total control of the environment by the community. It assumes the permanence and uniformity of our criteria of choice. The notion of continuous and unbroken progress is merely a slightly improved variant of the morality of the good. By coordinating their intelligences and their imaginations, members of intelligent communities promote the growth of the best; they create a best that is always new and always different. The best is continuously displaced not only because objective situations evolve, but because our understanding of situations develops or becomes confused (which also constitutes a changed situation), because the criteria of choice change as a function of the transformation of the environment and the evolution of our plans. Each new choice is seen from the point of view of an original and unforeseeable path of collective apprenticeship and self-invention.

The idea of collective intelligence does not defer the question of happiness. Far from implying any form of self-sacrifice, it encourages us to increase the degrees of freedom of individuals and groups, to implement win-win strategies, to create synergies between knowledge and knower. The collective intelligence has no enemies. It doesn't fight power; it abandons it. It doesn't pursue domination but germination, promotes the greatest possible variety of existents. The expansion of the powers of life and the qualities of being is the ultimate criterion—the most general, quasi-ontological—our choice of the best, represented by the flight from destruction and the tropism toward existence, an existence that includes not only facts but everything that is imaginable, imagined, possible, feasible.

Through collective intelligence we can continue the project of emancipation begun during the Enlightenment. Yet here too, we are unable to maintain the fiction of a linear, automatic, guaranteed progress. An extraordinary sense of sobriety about modernity has marked the end of the twentieth century. The archaic and barbarous are always present, ready to rise up, more archaic and barbarous than ever before. Everything coexists. Globalization (which turns all wars into civil wars) and nationalism, the

triumph of organized crime and the refinements of bio-
ethics, the cultural continent of urban youth with its en-
signs and music and child labor, widespread famine in
the face of a global megamachinery for the production of
dreams of interactive entertainment, high-tech multi-
nationals and the scarcity of water, cyberspace and illit-
eracy. Time is no longer linear; it is multiple, spiraling
around us. Perhaps we are not *post*modern, perhaps we
are not living after but *before* history, all temporality
blending together in a fabled moment, the source of a
history still to come that has not yet begun to flow. It
might be said that we are living in the "time of origins," in
the *arche* itself, within the time of myth, the great era of
metamorphoses and talking animals. Rhythms, spaces,
identities, possibilities are so many marks on the ivory
dice shaken by time. Not Chronos, the terrifying god who
ate his children and castrated his father, the god of linear
succession, but *aïon*, the time of time, eternity, innocence.
Time is a child playing dice. What kind of humanity will
his game produce? A world of planetary civil war rises up
before us, dominated by criminal networks and high-tech
elites, condemning the majority of humanity to a future of
hopeless misery. Collective intelligence provides an alternative.

Is it possible? Can we escape the struggle for power,
the efforts of domination, war? "War is the father and
king of all." Challenging Heraclitus's sentence, I appeal
to the judge of hell.

Several centuries before the flowering of the Greek
city states, the Hellenic world was dominated by Mycenaean civilization. In the *Iliad*, it is Agamemnon, king
of Mycenae, who leads the Achean expedition against
Troy. Today, when the traveler finds himself before the
ruins of the ancient fortress restored by archeologists, he
discovers that the walls are several meters thick, made of

enormous, cyclopean blocks of stone. In this warlike civilization all human effort, all material accumulation, was used to maintain the distinction between inside and outside.

Very different from the Mycenaean fortress—and much earlier—the palace of Knossos was for seven centuries the center of Minoan civilization. The Cretan palace is unfortified. The peaceful Minoan culture expended its energy in complex architecture, decorated halls, the beauty and ingeniousness of interior design (sewers, a system of drinking water, etc.). All the energy invested by the Mycenaeans in their ramparts was employed at Knossos in refining a way of life, complicating the design, the proliferation of a wealth of architectural detail: stairways, courtyards, columns, statues, floors, terraces, antechambers, large halls, small, secret rooms, treasuries, twisting passages, parapets, dead-ends.... The palace of Knossos is infinitely complex but its courtyards and light wells are open to the sky and sun; its windows and doors are open to the world and the city. Its paved roads lead to the other palaces in the great Cretan cities. Because they were not part of a warlike civilization, because their minds were directed toward problems other than defense and aggression, relationships of force and domination, the Minoans, at the same time as they interacted with other societies through art and commerce, folded and refolded their world upon itself, promoted the aesthetic wealth that preceded and possibly helped determine the "Greek miracle." In place of the rampart, the Minoans invented the labyrinth, cultural complexity, collective intelligence projected on architectural space.

Who is the minotaur? The terrifying beast that devoured young Athenians in its dark cave? This version of the minotaur legend is Greek. But the warlike Greeks, sons of Mycenae and readers of the *Iliad*, were unable to understand Knossos, the enigma of an irenic civilization. The minotaur, the man-bull, is none other than the Minoan acrobat who executes his dangerous rituals on the back of the sacred bull. The minotaur, the hybrid man-bull, stands at the center of the labyrinth, the central courtyard of the palace of Knossos. The creature moves about in the open air, light and graceful, in the sun-drenched square of the palace.

The Minoans were never conquered in war. Their culture collapsed following a series of natural catastrophes and migrations that led them far from their island home. No corpses have been found in the ruins of the burned palace. The mainland Greeks settled in Crete long after the decline of its original civilization.

Theseus killed the minotaur. The Mycenaeans overshadowed Minoan civilization, an artistic, technocentric civilization but one without weapons or slaves. The polemic Greeks buried irenic Crete. Beneath the conflict, lies peace. The Greeks buried Minos, deep beneath the earth, and made him the judge of hell. Beneath the transparent guise of Zeus, it is the Minoan bull who carries Europa.

Collective intelligence assumes that we will abandon the perspective of power. It strives to open the central void, the shaft of light that enables the interplay with alterity, chimerization, and labyrinthine complexity. But in the eyes of war, which is unable to acknowledge anything but itself, the palace of light, the white labyrinth, the joy, beauty, and sovereign lightness of architectural detail, becomes the black labyrinth, the mortal trap harboring the man-eating monster. The legend of the labyrinth represents our inability to find peaceful solutions. Extending both to the distant Cretan past and the opaque planetary future, the culture of strength and peace appears indecipherable. Linear B, the written script used by the Mycenaeans in Crete, has been decoded. But we still haven't found the key to Linear A, used by the Minoans before the Mycenaean conquest. The enigma of peace remains sealed. It awaits our decipherment of Linear A or, rather, the invention of a dynamic ideography, a writing of the future, the over-language of intelligent communities. Instead of reinforcing the fortresses of power, we must refine the architecture of cyberspace, the ultimate labyrinth. On each integrated circuit, on every microchip, we can discern but cannot decipher the secret key, the complicated emblem of collective intelligence, the irenic message scattered to the winds.

Notes

Foreword

[1] Jacques Ellul, *The Technological Society* (New York: Vintage Books, 1964).

[2] Simon Nora and Alain Minc, *The Computerization of Society: A Report to the President of France*, introduction by Daniel Bell (Cambridge, Massachusetts: MIT Press, 1980).

[3] See for example: Pierre Teilhard de Chardin, *The Future of Man*, translated from the French by Norman Denny. (New York: Harper & Row, 1969).

[4] For an analysis of McLuhan and his anticipation of contemporary developments in computing see: Eugene F. Provenzo, Jr., "Marshall McLuhan: Thirty Years Later," *Taboo: The Journal of Culture and Education*, Vol. 2 (Fall 1995): 206–211.

[5] See for example: Douglas Engelbart, "A Conceptual Framework for the Augmentation of Man's Intellect." In P. W. Howerton and D. C. Weeks, *Vistas in Information Handling: Vol. 1. The Augmentation of Man's Intellect by Machine* (Washington, D.C.: Spartan Books, 1963) 1–29. Engelbart published this essay while working at the Stanford Research Institute. In it, he outlined not only the basic principles of word-processing, but also the use of icon systems for computers, as well as technologies such as the computer mouse and scanning.

[6]*Ibid.*

[7]Engelbart's current work can be best understood by visiting The Bootstrap Institute online at: http://www2.bootstrap.org/services.htm.

[8]See Eugene F. Provenzo, Jr., *Beyond the Gutenberg Galaxy: Microcomputers and the Emergence of Post-Typographic Culture* (New York: Teachers College Press, 1986) and *Computing, Digital Culture and Pedagogy: The Analytical Engine* (New York: Peter Lang, in press).

Prologue

[1]See Howard Rheingold's excellent book, *The Virtual Community* (New York: Addison-Wesley, 1993), which traces the history of computer-mediated communication. Rheingold's book illustrates the different political, social, cultural, and ludic aspects involved, and clearly states the social risks associated with the development of the "electronic highway."

[2]See Josef Reichholf, *L'Émergence de l'homme: l'apparition de l'homme et ses rapports avec la nature* (Paris: Flammarion, 1991).

Introduction

[1]This hypothesis was inspired by the work of Bernard Perret. See *L'Économie contre la société. Affronter la crise de l'intégration culturelle et sociale,* by Bernard Perret and Guy Roustang (Paris: Editions du Seuil, 1993).

[2]This long-range approach to the "end of employment" was suggested by Robert Reich, *The Work of Nations: Preparing Ourselves for 21st Century Capitalism* (New York: Random House, 1991).

[3]The word was coined by Roy Ascott during the "Telenonia" conference held in Toulouse in 1992 as part of the FAUST project. See also, "Telenonia" in *Interactive Art, Intercommunication,* 7 (1994): 114–123, Tokyo, and "Telenonia, On Line" in *Kunst im Netz* (Graz: Steirischen Kulturinitiative, 1993), 135–146.

[4]I can't emphasize too strongly how much this vision owes to the Mutual Knowledge Exchange Network Movement (MRERS), directed by Claire and Marc Hébert-Suffrin. See, for example, their *Échanger les savoirs* (Paris: Desclée de Brouwer, 1991).

[5]Knowledge trees are a computer-based method for the overall management of skills in schools, businesses, job pools, local commu-

nities, and associations. This method is today being experimented with in several parts of Europe, especially France, in large corporations like Électricité de France and PSA (Peugeot and Citroën), mid-size companies, universities, business schools, local communities (municipalities in the Poitou-Charentes region of France), low-cost housing communities, etc.

Using this approach, each member of a community can spread awareness of the diversity of his skills, even those that are not validated by traditional scholarly and university systems. Growing from autobiographical information, a knowledge tree makes visible the organized multiplicity of skills available in a community. It is a kind of dynamic map, viewable on screen, and resembling a tree in appearance, each community giving rise to a differently shaped tree.

By skills I am referring to behavioral abilities (*knowledge-of-being*) as well as "know-how" or theoretical knowledge. Each elementary skill is recognized in individuals by means of a *brevet*, on the basis of a precisely specified procedure (tests, nomination by one's peers, the supply of proof of some sort, etc.). Visible on screen, the dynamic map of a group's know-how doesn't result in any *a priori* classification of knowledge: automatically produced by software, it is the expression, evolving in real time, of the trajectory of apprenticeship and experience of members of the community. The tree of a community grows and is transformed as that community's skills evolve. Thus, brevets for basic forms of knowledge will be located in the "trunk." Brevets for highly specialized forms of knowledge, associated with more extensive education, will form the "leaves." The "branches" will combine skills that are almost always associated with one another in the individual lists of individual skills, etc. But the organization of knowledge expressed by a tree is not permanent. It reflects the collective experience of a human group and thus evolves with that experience. For example, a brevet located on a leaf at a time t, can be lowered on a branch at time $t+n$. The tree, which is different for each community, doesn't reflect the customary partitions into disciplines, levels, successive phases of study, or institutional divisions. On the contrary, the mechanism for dynamic indexing and navigation that it offers produces a knowledge space without separations, one that is continuously reorganized depending on context and issue.

The representation as a knowledge tree means that simple inspection can be used to locate the position occupied by a given kind of knowledge at a given moment and the possible apprenticeship itineraries for reaching a given skill. Each individual has a personal image (an original distribution of brevets) on the tree, an image that he or she can refer to at any moment. This image is called the person's *blazon* to indicate that today's true nobility has been conferred on the basis of skill. Thus people obtain a better awareness of their situation in the "knowledge space" of the communities in which they participate and can consciously elaborate their own apprenticeship strategies.

Electronic messages, "routed on the basis of knowledge," correlate the supply and demand of know-how within the community and announce the availability of training and exchange for each elementary skill. Such an instrument thus serves the social bond through the exchange of knowledge and the employment of skills. All transactions and queries recorded by the software help continually determine the value (always contextual) of elementary skills as a function of different economic, pedagogic, and social criteria. This evaluation, extended through use, is an essential mechanism of self-regulation.

At the level of the community, the system of skill trees can contribute to the struggle against exclusion and unemployment by recognizing the know-how of those without diplomas, by favoring a better adaptation of training to employment, by stimulating a genuine "skill market." At the level of the network of schools and universities, the system can be used to implement a decompartmentalized and personalized cooperative pedagogy. Within an organization knowledge trees provide instruments for identifying and mobilizing know-how, the evaluation of training, as well as a strategic vision of skill evolution and need.

Enabling all types of learning methods to result in some qualification, the "tree" mechanism provides for improved skills management. Conversely, by evaluating the signs of skill in real time, skills management helps validate qualification. Every person who defines himself by obtaining a certain number of signs of competence also becomes accessible on the network. He is indexed to the navigation space and can be contacted for mutual exchanges of knowledge or skill requests. An improvement of the process of qualification thus has positive effects on sociability.

By responding to the speed of evolution and knowledge, this instrument makes visible in real time the rapid evolution of highly diverse skills. By allowing the diversity of skills to be expressed, it doesn't compartmentalize individuals within a job or category, thus promoting continuous personal development.

Today, every country has a diploma system and a structure for recognizing different types of knowledge. Moreover, within a given country, diplomas—notoriously lacking in this respect—are the only system of representation for skills shared by all branches of industry, all business and social environments. For the rest, the greatest heterogeneity is commonplace. The knowledge tree, however, can translate other systems of knowledge recognition and mutualize the signs of skill. These signs, or brevets, can be shared by every tree, even though they might be located and evaluated differently on each. Individuals maintain the same apprenticeship curriculum (individual list of brevets) when moving from one tree to another, and yet their image is different on each tree since it is projected against another collective "background."

[6]See *Les Arbres de connaissances* (preface by Michel Serres) by Michel Authier and Pierre Lévy (Paris: La Découverte, 1992). The current work is no more than a footnote to the knowledge trees currently taking shape in businesses, schools, and other organizations, and provides us with a concrete social and technical awareness of the nature of collective intelligence.

[7]The knowledge tree is an illustration of the technical and social feasibility of such a project.

Chapter 3

[1]See K. Eric Drexler and Chris Petersen, *Unbounding the Future: The Nanotechnology Revolution* (New York: William Morrow & Company, 1991).

Chapter 4

[1]On the question of government structures and the ability to govern, see the Club of Rome's report by Alexander King and Bertrand Schneider, *Questions de survie* (Paris: Calman-Lévy, 1991), 162–183, originally published in English as *The First Global Revolution*.

[2]The change in the technical and communications sphere, the impossibility of controlling our environment or using customary means of decision-making in the face of the flood of information, as well as the various pathologies associated with this new situation, have been thoroughly analyzed by Franco Berardi. See his *Mutazione e cyberpunk* (Genoa: Costa & Nolan, 1994), *Come si cura il nazi* (Rome: Castelvecchi, 1993), and Franco Berardi and Franco Bolleli, *Per una deriva felice* (Milan: Multipla, 1993).

[3]Ports in a digital interactive communications network.

[4]I am referring primarily to the *Critique de la raison politique* by Régis Debray (Paris: Gallimard, 1981), whose views on politics were recently collected in *Manifestes médiologiques* (Paris: Gallimard, 1994). See also, by the same author, *Vie et mort de l'image* (Paris: Gallimard, 1992), *L'État séducteur* (Paris: Gallimard, 1993), and *Cours de médiologie générale* (Paris: Gallimard, 1991). The notion of heteronomy as an essential component in politics considerably weakens the author's "mediologic" thesis. There is nothing to prevent heteronomy, like other aspects of community life, from being dependent on the state of contemporary communications technologies and practices. Why limit the dichotomy heteronomy/autonomy to the field of mediology? The desire to construct an anthropology around a stable position devoid of the technological and linguistic *becoming* that is the very essence of the human is a recipe for failure.

[5]See Alvin Toffler, *Powershift: Knowledge, Wealth, and Violence at the Edge of the 21st Century*, and Alvin and Heidi Toffler, *War and Anti-War: Survival at the Dawn of the Twenty-First Century* (Boston: Little Brown, 1993).

Chapter 5

[1]Principal sources were Henri Corbin, *Histoire de la philosophie islamique*, vol. 1, *Des origines jusqu'à la mort d'Averroès* (Paris: Gallimard, 1964), especially the sections in chap. V ("Les philosophes hellenisants") devoted to Al-Fārābī, Ibn Sina, Abū'l-Barakāt al-Baghdādī, and Maimonides, *Le Guide des égarés* [*The Guide for the Perplexed*], translated by Salomon Munk (Lagrasse: Verdier, 1979).

[2]See Pierre Lévy "Le cosmos pense en nous," in *Chimères*, 14 (1992): 63–79, reprinted in *Les Nouveaux outils de la pensée*, edited by Pierre Chambat and Pierre Lévy (Paris: Éditions Descartes, 1992).

[3]In Maimonides there are only four heavens and four separate intelligences. The agent intellect is, therefore, the Fourth Cherub rather than the Tenth. The number of heavens and separate intelligences (ten, seven, or four depending on the author) is not only related to mystical and theological speculation, but is based on astronomical considerations as well. The process of emanation cannot continue indefinitely because the force of the divine influx is not infinite. (See my previous comments at the beginning of this chapter on the "hatred of the infinite" found in the Platonic and Aristotelian traditions.)

[4]"Agent intellect is to man's possible intellect what the sun is to the eye, which remains potential vision as long as it remains in shadow" (al-Fārābī).

[5]In addition to the work of Henri Corbin, previously mentioned, see *Le Guide des égarés* by Maimonides, pt. II, chap. 4.

[6]See *Le Guide des égarés*, pt. II, chap. 37.

[7]Autopoiesis is the process of continuous self-renewal. The concept of autopoiesis has received great attention in theoretical biology, notably by Humberto Maturana and Francisco Varela. See Francisco Varela, *Principles of Biological Autonomy* (North Holland, NY: Elsevier, 1980).

[8]Can we ever escape the problem of power? For a discussion of this problem, see the epilogue of the present book, entitled "Voyage to Knossos."

[9]The remainder of this chapter is largely based on the chapter, "L'intellect, l'intelligible, l'intelligent," in Maimonides's *Le Guide des égarés*, which is itself the culmination of a long, rich tradition of thought that flowered during the great epoch of Islamic civilization.

[10]See Pierre Lévy, *Les Technologies de l'intelligence* (Paris: La Découverte, 1990, reissued in the collection Points-Sciences by Éditions de Seuil, 1993), in which I present a detailed argument for this point of view along with an extensive bibliography on the anthropology and psychology of cognitive processes that have been "augmented" by sign systems, instruments of representation, and communications tools.

[11]See Pierre Lévy, *L'Idéographie dynamique, vers une imagination artificielle* (Paris: La Découverte, 1991), which describes the technical, cognitive, and linguistic conditions for an interactive, digitally-based cinelanguage.

Chapter 6

[1]William Gibson, *Neuromancer* (New York: Ace Science Fiction Books, 1984).

Chapter 8

[1]Michel Serres, "J'habite une multiplicité d'espaces," in *L'Interférence* (Paris: Éditions de Minuit, 1972), 151.

[2]Pierre Clastres, *La Société contre l'état, essais d'anthropologie politique* (Paris: Éditions de Minuit, 1974).

[3]Gilles Deleuze and Félix Guattari, *L'Anti-Œdipe* (Paris: Éditions de Minuit, 1972), especially the chapter entitled "Sauvages, barbares, civilisés."

Chapter 9

[1]In *L'Anti-Œdipe*, Gilles Deleuze and Félix Guattari demonstrate the close relationship between capitalism and a subjectivity that is narrowly articulated around the family. See, for example, the chapter entitled "Sauvages, barbares, civilisés."

[2]See Michel Authier and Pierre Lévy, "La cosmopédie, une utopie hypervisuelle," in *Culture technique*, 24 (April 1992): 236–244, in an issue devoted to "Machines à communiquer."

[3]See Michael Authier and Pierre Lévy, *Les Arbres de connaissances*, op. cit.

[4]The blazon is the image of an individual's apprenticeship (*curriculum*) as it appears on the knowledge tree of a community. See *Les Arbres de connaissances*, op. cit.

[5]See Francisco Varela, *Principles of Biological Autonomy*, op. cit., Humberto Maturana and Francisco Varela, *The Tree of Knowledge* (Boston: New Science Library, 1987). See also my "Analyse de contenu des travaux du Biological Computer Laboratory," in *Cahiers du CREA* 8 (1986): 155–191, which can be obtained from CREA, 1, rue Descartes, 75005 Paris, France.

[6]With respect to the concept of intellectual technology, see Pierre Lévy, *Les Technologies de l'intelligence*, op. cit.

Chapter 10

[1]Charles Baudelaire, "Correspondences," in *Les Fleurs du mal*, 1857.

[2]Daniel Bougnous, *La Communication par la bande* (Paris: La Découverte, 1991).

[3]See François Rastier's remarkable study, "La Triade sémiotique, le trivium et la sémantique linguistique," *Nouveaux Actes Sémiotiques* 9 (1990): 54.

[4]See Mikhail Bakhtine, *Le Marxisme et al philosophie du langage* (Paris: Éditions de Minuit, 1977).

[5]Guy Debord, *La Société du spectacle* (Paris: Buchet-Chastel, 1967, reprinted by Gallimard in 1993).

[6]Cf. Jean-Pierre Balpe, *Les Hyperdocuments* (Paris: Eyrolles, 1990), René Laufer and Domenico Scavetta, *Les Hypertextes* (Paris: PUF, 1992), and Pierre Lévy, *Les Technologies de l'intelligence*, op. cit.

[7]Cf. Howard Rheingold, *Virtual Reality*, 1991.

[8]Cf. *Artificial Life* (New York: Addison-Wesley, 1989), ed. Christopher G. Langton, and *Artificial Life 2*, eds. Christopher G. Langton, Charles Taylor, J. Doyne Farmer, and Steen Rasmussen (New York: Addison-Wesley, 1992). These two books are based on international colloquia organized by the Santa Fe Institute for the Sciences of Complexity.

[9]Cf. Pierre Lévy, *L'Idéographie dynamique*, op. cit.

[10]Cf. Michel Authier and Pierre Lévy, *Les Arbres de connaissance*, op. cit.

Chapter 11

[1]Jean-Jacques Rousseau, *Discours sur l'inégalité*, pt. 2, 1754.

[2]See Michel Serres, *Rome, le livre des fondations* (Paris: Grasset, 1983).

[3]And more recently with the period of the radiation given off by the cesium-133 isotope.

[4]In Charles Chaplin's celebrated film "Modern Times" (1935), the worker (Chaplin) is martyred by the machines and assembly lines in a factory and nearly crucified on the hands of a giant clock.

[5]"There are many periods in our life. There are many streams blended in time." Christian Bobin, *La Part manquante* (Paris: Gallimard, 1992), 29.

[6]Michel Authier and Pierre Lévy, *Les Arbres de connaissances*, op. cit.

[7]Norbert Wiener, *Cybernetics* (New York: MIT Press, 1961).

[8]The many barriers to knowledge obviously exist because our schools, universities, and professional disciplines are structured like territories.

Chapter 12

[1]See Pierre Lévy, *De la programmation considérée comme un des beaux-arts* (Paris: La Découverte, 1992), and especially pt. 1, "Jeux d'aventures."

[2]On the concept of the point as opposed to the vanishing line, see Gilles Deleuze and Félix Guattari, *Mille plateaux* (Paris: Éditions de Minuit, 1980).

[3]On the history of statistics, see Alain Desrosières, *La Politique des*

grands nombres. Histoire de la raison statistique (Paris: La Découverte, 1993).

[4]On the concept of dynamic configuration, see the remarkable work by François Julien, *La Propension des choses, une histoire de l'efficacité en Chine* (Paris: Éditions de Seuil, 1992).

Chapter 13

[1]Michel Serres, *Les Origines de la géométrie* (Paris: Flammarion, 1993).

[2]Michel Serres, *Hermès IV. La distribution* (Paris: Éditions de Minuit, 1977).

[3]Fundamental texts on the subject are: Claude Shannon, *The Mathematical Theory of Communication* (University of Illinois Press, 1949) and Norbert Wiener, *Cybernetics*, op. cit. *Cybernetics and Society*, also by Wiener, contains an excellent summary of these ideas and provides the basis for a global vision of society.

[4]Henri Atlan, *Entre le cristal et la fumée* (Paris: Éditions du Seuil, 1979) and *L'Organisation biologique et la théorie de l'information* (Paris: Hermann, 1972).

[5]For a more detailed analysis of these issues, see chaps. 4, 5, and 6 of *La Machine univers* (Paris: La Découverte, 1987).

[6]See Ilya Prigogine and Isabelle Stengers, *La Nouvelle alliance* (Paris: Gallimard, 1979) and *Entre le temps et l'éternité* (Paris: Fayard, 1988).

[7]See James Gleick, *Chaos: Making a New Science* (New York: Viking Press, 1987).

[8]Benoît Mandelbrot, *Les Objets fractals* (Paris: Flammarion, 1975). It is worth noting that the first studies of fractal objects involved stock-market fluctuations.

[9]Michel Authier and Pierre Lévy, *Les Arbres de connaissances*, op. cit.

[10]See Pierre Lévy, "L'univers aleph, pour une cinécartographie de l'information," in *Actes de la sixième école d'automne du Campus Thomson*, September 1992, 44 pages.

Chapter 14

[1]Michel Serres, *Les Origines de la géométrie*, op cit.

[2]Thomas Kuhn, *The Structure of Scientific Revolutions*. For an excellent critical discussion of the most important contemporary epistemological theories, see Isabelle Stengers, *L'Invention des sciences mod-*

ernes (Paris: La Découverte, 1993). Stengers's book contains one of the most sophisticated approaches to the ethics of science (and knowledge in general) yet to be formulated.

[3] The epistemology (or anti-epistemology) of technoscience has been described by Bruno Latour and others working in the new field of the anthropology of science and technology. See, for example, Bruno Latour, *La Science en action* (Paris: La Découverte, 1989), and, under the direction of Michel Callon, *La Science et ses réseaux* (Paris: La Découverte, 1989). Concerning the problem of circulation in contemporary science, one excellent source is Charles Halary, *Les Exilés du savoir. Les migrations scientifiques internationales et leurs mobiles* (Paris: L'Harmattan, 1994).

[4] Michel Serres, *Le Système de Leibnitz et ses modèles mathématiques* (Paris: PUF, 1968). See also Serres's *La Traduction* (Paris: Éditions de Minuit, 1972) and *L'Interférence*, op. cit.

[5] On computerized simulations as instruments of computer-assisted imagination, see Pierre Lévy, *Les Technologies de l'intelligence*, op. cit., especially chaps. 10 and 13, and *L'Idéographie dynamique*, op. cit.

[6] For additional details on the cosmopedia, see Michel Authier and Pierre Lévy, "La cosmopédie, une utopie hypervisuelle," op. cit.

[7] On the concept of the subject creating a subject by its own initiation, or autopoiesis, see Francisco Varela, *Principles of Biological Autonomy*, op. cit.

[8] See Pierre Lévy, "Le cosmos pense en nous," op. cit.

[9] On the construction of the object through the implication of subjectivities, see Michel Serres, *Statues* (Paris: François Bourin, 1990).

[10] On the important concept of heterogenesis, see Félix Guattari, *Chaosmose* (Paris: Gallimard, 1992). Heterogenesis, or the creation of alterity, should be compared to homogenesis or homogenization (think of homogenized milk), which consists of creating or manufacturing something similar to oneself, something homogeneous.

Chapter 15

[1] See Cornelius Castoriadis, *Domaines de l'homme, les carrefours du labyrinthe*, II (Paris: Éditions du Seuil, 1986) and *Le Monde morcelé, les carrefours du labyrinthe*, III, 1990.

[2] See Michel Authier and Pierre Lévy, *Les Arbres de connaissances*, op. cit.

[3] See Félix Guattari, *La Révolution moléculaire* (Paris: 10/18, 1980).

[4]Heraclitus.

[5]Charles Halary, *Les Exilés du savoir. Les migrations scientifiques internationales et leurs mobiles*, op. cit.

[6]See Michel Authier and Rémi Hess, *L'Analyse institutionnelle* (Paris: PUF, 1981) and Michel Serres, *Le Parasite* (Paris: Grasset, 1980).

[7]See Michel Authier and Marie Thonion, *Secret et sécurité*, SPES report, France Télécom, 1982.

[8]"Thirty spokes converge on the hub, but it is the central void that causes the wagon to advance." Lao-tzu, *Tao Te Ching*, chap. xi.

Index